Mindf*ck

Inside Cambridge Analytica's Plot to Break the World

CHRISTOPHER WYLIE

PROFILE BOOKS

This paperback edition published in 2020

First published in Great Britain in 2019 by
Profile Books Ltd
29 Cloth Fair
London
EC1A 7NN

www.profilebooks.com

Published in the United States of America in 2019 by
Random House, an imprint and division of Penguin Random House LLC, New York

3 5 7 9 10 8 6 4 2

Printed and bound in Great Britain by
CPI Group (UK) Ltd, Croydon, CR0 4YY

A CIP catalogue record for this book is available from the British Library.

ISBN 978 1 78816 500 6
Export edn 978 178816 506 8
eISBN 978 1 78283 677 3
Audio ISBN 978 1 78283 715 2

To my parents, Kevin and Joan,
who taught me to be brave,
to stand up for myself,
and to do the right thing

On résiste à l'invasion des armées; on ne résiste pas à l'invasion des idées.

(One withstands the invasion of armies; one does not withstand the invasion of ideas.)

— VICTOR HUGO

CONTENTS

Mindf*ck

1

GENESIS

WITH EACH STEP, MY NEW SHOES DIG INTO MY HEELS. I CLUTCH a dark-blue binder, filled with documents organised by coloured tabs. Awestruck by where I've found myself, and apprehensive of where I'm heading, I focus on the sounds of our footsteps. An aide reminds us to move quickly so we won't be seen. We walk past uniformed guards, into an atrium, and turn down a corridor. The aide pushes open a door and we rush down some stairs and into a hallway that looks exactly like the last one – marble floors, high ceilings, wooden doors with the occasional American flag. There are seven of us, and our footsteps echo through the hall. We are close; then I'm caught. A congressman spots me and waves hello. *Back again already?* A handful of journalists wander out of a press conference. They clock my electric pink hair and know who I am.

Two cameramen run in front of me and start filming, walking backward as they do. A scrum forms, the questions start coming – *Mr Wylie, a question from NBC! A question from CNN! Why are you here?* – and one of my lawyers reminds me to keep my mouth shut. The aide points me to a lift, warning the journalists to keep their distance, and we pile in. The cameras keep snapping as the doors close.

I'm jammed in the back of the lift, surrounded by people in suits. We start to descend, dropping deep underground. Everyone stays quiet on the way down. My mind is swimming with all of the prep work I've done with my lawyers – what US laws were broken and by whom, what rights I do and don't have as a non-citizen visiting America, how to calmly respond to accusations, what happens if I am arrested afterward. I have no idea what to expect. No one does.

We come to a stop and the elevator doors glide open. There's nothing down here except another door, with a large red sign that reads RESTRICTED AREA in white lettering. NO PUBLIC OR MEDIA ACCESS. We're three floors beneath the US Capitol, in Washington, DC.

Beyond the door, the floors are covered in a plush maroon carpet. Uniformed guards confiscate our phones and other electronics, placing them on a numbered shelf behind the desk, one to a person, and giving us each a numbered ticket. They tell us we can have only pencils and paper beyond this point. And on the way out, they warn us, our papers could be confiscated if it's determined that we've taken notes on anything of a sensitive nature.

Two guards push open a massive steel door. One of them gestures us through, and one by one we step into a long hallway dimly illuminated by fluorescent lights. The walls are panelled in dark wood, and the corridor is lined with long rows of American flags on stands. It smells like an old building, stale and musty, with hints of cleaning fluid. The guards lead us down the hall, turning left and continuing to yet another door. Above, a wooden seal emblazoned with a giant eagle, arrows clutched in its talons, stares down at us. We have arrived at our destination: the Sensitive Compartmentalised Information Facility (SCIF) of the United States House Permanent Select Committee on Intelligence – the same room where classified congressional briefings are held.

Inside, hit by the glare of fluorescent lights, my eyes need time to adjust. The space is thoroughly nondescript, with blank beige walls and a conference table surrounded by chairs. It could be any room in any of the numerous bland federal buildings scattered across Washington, but I'm struck by the silence of the SCIF. It is soundproof, built with multilayer walls that make it impervious to surveillance. The architecture is said to be blast-proof. This is a secure space, a place for America's secrets.

Once we've taken our seats, the members of Congress begin filing in. Aides place tabulated binders on the table in front of each committee member – the Democrats' ranking member, California congressman Adam Schiff, sits directly across from me, and to his left sits Congresswoman Terri Sewell, with Eric Swalwell and Joaquin Castro clustered together at the far end. I'm flanked by my lawyers and my friend Shahmir Sanni, a fellow whistleblower. We give the Republicans a few minutes to show up. They never do.

It's June 2018, and I'm in Washington to testify to the US Congress about Cambridge Analytica, a military contractor and psychological warfare firm where I used to work, and a complex web involving Facebook, Russia, WikiLeaks, the Trump campaign and the Brexit referendum. As the former director of research, I've brought with me evidence of how Facebook's data was weaponised by the firm, and how the systems they built left millions of Americans vulnerable to the propaganda operations of hostile foreign states. Schiff leads the questioning. A former federal prosecutor, he is sharp and precise with his lines of inquiry, and he wastes no time getting to the heart of the matter.

Did you work with Steve Bannon? *Yes.*
Did Cambridge Analytica have any contacts with potential Russian agents? *Yes.*
Do you believe that this data was used to sway the American electorate to elect the president of the United States? *Yes.*

An hour goes by, then two, then three. I chose to come here of my own accord and to answer these questions about how a liberal, gay twenty-four-year-old Canadian found himself part of a British military contractor developing psychological warfare tools for the American alt-right. Fresh out of university, I had taken a job at a London firm called SCL Group, which was supplying the UK Ministry of Defence and NATO armies with expertise in information operations. After western militaries were grappling with how to tackle radicalisation online, the firm wanted me to help build a team of data scientists to create new tools to identify and combat extremism online. It was fascinating, challenging and exciting all at once. We were about to break new ground for the cyber defences of Britain, America and their allies and confront bubbling insurgencies of radical extremism with data, algorithms and targeted narratives online. But through a chain of events that unfolded in 2014, a billionaire acquired our project in order to build his own radicalised insurgency in America. Cambridge Analytica, a company few had ever heard of, a company that weaponised research in psychological profiling, managed to turn the world upside down.

In the military, when weapons fall into the wrong hands, they call it blowback. It looked as if this blowback had detonated in the White House itself. I could not continue working on something so corrosive to our societies, so I blew the whistle, reported the whole thing to the authorities, and worked with journalists to warn the public about what was going on. Sitting before this panel, jet-lagged from a transatlantic flight the day before, I still cannot help but feel on the spot as the questions grow more pointed. But several times, my attempts to explain the intricacies of the company's operations leave everyone with puzzled faces, so I simply pull out a binder and slide it to the congressmen. *What the hell,* I think. I've come this far, so I might as well give them everything I have with me. There is no break, and the door behind me remains closed the entire time. I'm locked in a stuffy, windowless room deep underground, with nowhere to look except straight into the eyes of these members of Congress as they all try to figure out what the hell just happened to their country.

THREE MONTHS BEFORE THIS, on 17 March, 2018, the *Guardian, The New York Times* and Channel 4 News had simultaneously published the results of a year-long joint investigation, spurred by my decision to reveal the truth about what was happening inside Cambridge Analytica and Facebook. My coming out as a whistleblower prompted the largest data crime investigation in history. In Britain, the National Crime Agency (NCA), MI5 (the UK's domestic intelligence agency), the Information Commissioner's Office, the Electoral Commission and London's Metropolitan Police Service all got involved. In the United States, the FBI, the Department of Justice, the Securities and Exchange Commission (SEC), and the Federal Trade Commission (FTC) jumped in.

In the weeks before that first story, the investigation by special counsel Robert Mueller had been heating up. In February, Mueller indicted thirteen Russian citizens and three Russian companies, charging them with two separate counts of conspiracy. A week later came indictments of former Trump campaign manager Paul Manafort and his deputy, Rick Gates. On 16 March, Attorney General Jeff Sessions fired FBI deputy director Andrew McCabe, just a little more than twenty-four hours before he was to retire with a pension. People were desperate for information about what had happened between the Trump campaign and Russia, but no one had been able to connect the dots. I provided evidence tying Cambridge Analytica to Donald Trump, Facebook, Russian intelligence, international hackers and Brexit. This evidence revealed how both an obscure foreign contractor engaged in illegal activity and the same foreign contractor had been used by the winning Trump and Brexit campaigns. The email chains, internal memos, invoices, bank transfer records and project documentation I brought demonstrated that Trump and Brexit had deployed the same strategies, powered by the same technologies, directed by many of the same people – all under the spectre of covert Russian involvement.

Two days after the story's release, an urgent question was brought to the British Parliament. In a rare moment of solidarity, government ministers and senior opposition members of Parliament sang as a unified chorus about Facebook's negligence in failing to prevent its platform from becoming a hostile propaganda network for elections and the implications for western democracies. The next wave of stories focused on Brexit, with the integrity of the referendum vote called into question. A collection of documents I provided to law enforcement revealed that the Vote Leave campaign had used secret Cambridge Analytica subsidiaries to spend dark money to propagate disinformation on Facebook and Google ad networks. This was determined to be illegal by the UK's Electoral Commission, with the scheme ending up as one of the largest and most consequential breaches of campaign finance law in British history. The office of 10 Downing Street descended into communication crisis as the evidence of Vote Leave's cheating emerged. The NCA and MI5 were later handed evidence of the Russian embassy's direct relationship with the largest funders of pro-Brexit campaigns during the referendum. A week later, Facebook's stock plummeted 18 per cent, amounting to an $80 billion loss in valuation. The turbulence would continue, culminating in what still stands as the largest single-day loss in share value in US corporate history.

On 27 March, 2018, I was called before Parliament for a live public hearing – something I'd get quite used to over the next several months. We covered everything from Cambridge Analytica's reliance on hackers and bribes to Facebook's data breach to Russian intelligence operations. After the hearing, the FBI, DOJ, SEC and FTC launched investigations. The US House Intelligence Committee, House Judiciary Committee, Senate Intelligence Committee and Senate Judiciary Committee all wanted to talk to me. Within weeks, the European Union and more than twenty countries had opened up inquiries into Facebook, social media and disinformation.

I told my story to the world, and now every screen was a mirror reflecting it back at me. For two weeks straight, my life was chaos. Days would start with appearances on British breakfast shows and European networks at 6 a.m. London time, continuing with interviews on US networks until midnight. Reporters followed me everywhere. I started to receive threats. Fearing for my safety, I had to hire bodyguards to protect me at public events. My parents, both physicians, had to temporarily close their medical clinic due to a frenzy of journalists asking questions and scaring patients. In the months that followed, my life became almost unmanageable, but I knew I had to keep sounding the alarm.

The story of Cambridge Analytica shows how our identities and behaviour have become commodities in the high-stakes data trade. The companies that control the flow of information are among the most powerful in the world; the algorithms they've designed in secret are shaping minds in ways previously unimaginable. No matter what issue you care about most – gun violence, immigration, free speech, religious freedom – you can't escape Silicon Valley, the new epicentre of our crisis of perception. My work with Cambridge Analytica exposed the dark side of tech innovation. We innovated. The alt-right innovated. Russia innovated. And Facebook, that same site where you share your party invites and baby pictures, allowed those innovations to be unleashed.

I SUSPECT I WOULDN'T have been interested in technology, or ended up at Cambridge Analytica, had I been born into a different body. I defaulted to computers because there was not much else available to a kid like me. I grew up on Vancouver Island, on the west coast of British Columbia, surrounded by oceans, forests and farmland. My parents were both doctors, and I was their eldest, followed by my two baby sisters, Jaimie and Lauren. When I was eleven, I started to notice that my legs were becoming stiffer and stiffer. I couldn't run as fast as the other kids, and I started to walk funny, which of course made me a target for bullies. I was diagnosed with two relatively rare conditions, whose symptoms included severe neuropathic pain, muscle weakness and vision and hearing impairment. By twelve I was in a wheelchair – just in time for the onset of adolescence – and I used that chair for the rest of my school days.

When you are in a wheelchair, people treat you differently. You can sometimes feel more like an object than a person – your means of getting around is how people come to understand and define you. You have to approach buildings and structures differently – *What entrances can I go in? How do I reach my destination while avoiding stairs?* You learn to look for things that other people never notice.

Not long after I discovered the computer lab, it became the one room at school where I didn't feel alienated. Outside, there were either bullies or patronising staff. Even when teachers shepherded other kids to interact with me, it was always done out of obligation, which became even more annoying than being ignored. Instead I'd go to the computer lab.

I started making webpages around age thirteen. My first website was a Flash animation of the Pink Panther being chased by a bumbling Inspector Clouseau. Soon after, I saw a video about programming Noughts and Crosses in JavaScript and thought it was the coolest thing ever. The game seems simple enough until you start having to break down all of the logic. You can't just let the computer randomly select a box, as that would be boring. You have to guide the computer with rules, like putting an *X* in a box adjacent to another *X* – that is, unless there is an *O* already in that row or column. And what about diagonal *X*s – how do we explain them?

Eventually I strung together hundreds of lines of spaghetti code. I still remember the feeling of making a move and then watching my little creation play back. I felt like a conjurer. And the more I practiced my incantations, the more powerful my magic could become.

Outside of computer lab, school remained an education in what I wasn't able or allowed to do, and who I could not be. My parents encouraged me to keep trying to find a place where I could fit in, so when I was fifteen, I spent the summer of 2005 boarding at Lester B. Pearson United World College, an international school in Victoria named after the Nobel Peace Prize-winning Canadian prime minister who conceptualised the world's first UN peacekeeping force during the 1950s Suez Crisis. Spending so much time with students from every part of the world was enthralling, and for the first time, I was actually interested in the lessons and what my peers had to say. I became friends with a survivor of the Rwandan genocide, who told me one evening when we were up late in our residence hall about how his family was murdered and what it was like walking alone all the way to a refugee camp in Uganda when he was just a child.

But it was after sitting at a dinner one evening in the dining hall where Palestinian and Arab students sat directly across from Israeli students, forcefully debating the future of their homelands, that I really started to wake up to the world around me. I realised how much I didn't know about what was happening – but I wanted to – and so I very quickly developed an interest in politics. The following school year, I began skipping class to attend town hall events with local members of Parliament. At school, I rarely talked to anyone, but at these events I felt free to express myself. In a classroom, you sit in the back while the teacher tells you how and what to think. There is a curriculum, a prescription of thought. But in a town hall, I discovered the opposite. Sure, the politician stands up front, but it is the people in the audience – *us* – who get to tell him or her what *we* think. That inversion was so incredibly appealing to me, and whenever members of Parliament would announce an event, I'd show up, ask questions, and even tell them what I thought.

It was liberating to find my voice. Like any teenager, I was exploring who I was, but for someone gay and in a wheelchair, this was an even bigger challenge. When I started attending these public forums, I began to realise that many of the things I was living through were not simply personal issues – they were also political issues. *My challenges were political. My life was political. My mere existence was political.* And so I decided to become *political*. An adviser to one of the MPs, a former software engineer named Jeff Silvester, took notice of this outspoken kid who always showed up. He offered to help me find a role in the Liberal Party of Canada (LPC), which was looking for tech help. Soon it was agreed, at the end of that summer I would start my first real job, as a political assistant at Parliament in Ottawa.

I spent the summer of 2007 in Montréal, hanging out in hacker spaces frequented by French Canadian techno-anarchists. They tended to gather in converted industrial buildings with concrete floors and plywood walls, in rooms decorated with retro tech like Apple IIs and Commodore 64s. By then, with treatment, I could shuffle around without a wheelchair. (I have continued to improve, but my physical limits were tested by my experience as a whistleblower. Just before the first Cambridge Analytica story was published, I had a seizure and collapsed, unconscious, on a South London pavement before waking up at University College Hospital to the sharp pain of a nurse inserting an IV needle into my arm.) Most hackers couldn't care less what you look like or if you walk funny. They share your love of the craft and want to help you get better at it.

My brief exposure to hacking communities left a permanent impression. You learn that no system is absolute. Nothing is impenetrable, and barriers are a dare. The hacker philosophy taught me that if you shift your perspective on any system – a computer, a network, even society – you may discover flaws and vulnerabilities. As a gay kid in a wheelchair, I came to understand systems of power early on in life. But as a hacker, I learned that every system has weaknesses waiting to be exploited.

SHORTLY AFTER I STARTED my job at the Canadian Parliament, the Liberal Party took an interest in what was happening down south. At that time, Facebook was just becoming mainstream and Twitter had just started gaining momentum; no one had any concept of how to use social media to campaign, because social media was in its infancy. But a rising star in US presidential politics was about to hit the accelerator.

While other candidates were twiddling their thumbs trying to figure out the internet, Barack Obama's team set up My.BarackObama .com and started a grassroots revolution. While other sites (like Hillary Clinton's) focused on putting up standard political advertisements, Obama's website centered on providing a platform for grassroots organisations to organise and execute get-out-the-vote campaigns. His website ratcheted up excitement around the Illinois senator, who was much younger and more tech savvy than his opponents. Obama felt like what a leader is supposed to be. And after spending my formative years being told about my limits, the defiant optimism he instilled in that simple message of *Yes, we can!* spoke to me. Obama and his team were transforming politics, so when I was eighteen, I was among several people sent to the United States by the Liberal Party to observe different facets of his campaign and identify new tactics that could be transported back for progressive campaigns in Canada.

At first, I toured a couple of early primary states, starting with New Hampshire, where I spent time talking to voters and seeing up close what American culture was really like. This was both fun and eye-opening; coming from Canada, I was struck by how different our sensibilities were. The first time an American told me he was dead set against 'socialised medicine', the same kind of public healthcare I accessed almost every month back home, I was shocked that someone could even think this way. The hundredth time, not so much.

I liked roaming around and talking with people, so when it was time to switch focus to the data group, I wasn't terribly excited to do it. But then I was introduced to Obama's national director of targeting, Ken Strasma, who quickly changed my mind.

The sexy part of the Obama campaign was its branding and use of new media like YouTube. This was the cool stuff, the visual strategy nobody had used before because YouTube was still so new. That was what I wanted to see, until Ken stopped me short. *Forget the videos,* he told me. I needed to go deeper, into the heart of the campaign's tech strategy. *Everything we do,* he said, *is predicated on understanding exactly who we need to talk to, and on which issues.*

In other words, the backbone of the Obama campaign was data. And the most important work Strasma's team produced was the modelling they used to analyse and understand that data, which allowed them to translate it into an applied fit – to determine a real-world communications strategy through ... artificial intelligence. *Wait – AI for campaigns?* It seemed vividly futuristic, as if they were building a robot that could devour reams of information about voters, then spit out targeting criteria. That information then travelled all the way up to the senior levels of the campaign, where it was used to determine key messages and branding for Obama.

The infrastructure for processing all this data came from a company then called the Voter Activation Network, Inc. (VAN), which was run by a fabulous gay couple from the Boston area, Mark Sullivan and Jim St George. By the end of the 2008 campaign, thanks to VAN, the Democratic National Committee would have ten times more data on voters than it had after the 2004 campaign. This volume of data, and the tools to organise and manipulate it, gave Democrats a clear advantage in driving voters to the polls.

The more I learned about the Obama machine, the more fascinated I was. And I later got to ask all the questions I wanted of Mark and Jim, as they seemed to find it amusing that this young Canadian had come to America to learn about data and politics. Before I saw what Ken, Mark and Jim were doing, I hadn't thought about using maths and AI to power a political campaign. In fact, when I first saw people lined up at computers at the Obama headquarters, I thought, *Messages and emotions, not computers and numbers, are what create a winning campaign.* But I learned that it was those numbers – and the predictive algorithms they created – that separated Obama from anyone who had ever run for president before.

As soon as I realised how effectively the Obama campaign was using algorithms to deliver its messages, I started studying how to create them on my own. I taught myself how to use basic software packages like MATLAB and SPSS, which let me mess around with data. Instead of relying on a textbook, I started by playing with the

Iris data set – the classic data set for learning statistics – and learned by trial and error. Being able to manipulate the data, which involved using the different features of irises, like petal length and colour, to predict species of flowers, was absolutely absorbing.

Once I understood the basics, I switched from petals to people. VAN was filled with information on age, gender, income, race, homeownership – even magazine subscriptions and airline miles. With the right data inputs, you could start to predict whether people would vote for Democrats or Republicans. You could identify and isolate the issues that were likely to be most important to them. You could begin to craft messages that would have a greater chance of swaying their opinions.

For me, this was a wholly new way of understanding elections. Data was a force for good, powering this campaign of change. It was being used to produce first-time voters, to reach people who felt left out. The deeper I got into it, the more I thought that data would be the saviour of politics. I couldn't wait to get back to Canada and share with the Liberal Party what I'd learned from the next president of the United States.

In November, Obama achieved a decisive victory over John McCain. Two months later, after friends in the campaign extended an invitation to the inauguration, I flew to Washington to party with the Democratic victors. (First came a slight kerfuffle at the door, when staff freaked out about letting the under-twenty-one me into the open-bar event.) I had an incredible evening, chatting with Jennifer Lopez and Marc Anthony, watching Barack and Michelle Obama enjoy their first dance as the First Couple. A new era had dawned, and now came a chance to celebrate what could happen when the right people understood how to use data to win modern elections.

BUT BY DIRECTLY COMMUNICATING select messages to select voters, the microtargeting of the Obama campaign had started a journey toward the privatisation of public discourse in America. Although direct mail had long been part of American campaigns, data-driven microtargeting allowed campaigns to match a myriad of granular narratives to granular universes of voters – your neighbour might receive a wholly different message than you did, with neither of you being the wiser. When campaigns were conducted in private, the scrutiny of debate and publicity could be avoided. The town square, the very foundation of American democracy, was incrementally being replaced by online ad networks. And without any scrutiny, campaign messages no longer even had to look like campaign messages. Social media created a new environment where campaigns could now appear, as Obama's campaign piloted, as if your friend was sending you a message, without your realising the source or calculated intent of that contact. A campaign could look like a news site, university or public agency. With the ascendancy of social media, we have been forced to place our trust in political campaigns to be honest, because if lies are told, we may never notice. There is no one there to correct the record inside of a private ad network.

In the years leading up to the first Obama campaign, a new logic of accumulation emerged in the boardrooms of Silicon Valley: Tech companies began making money from their ability to map out and organise information. At the core of this model was an essential asymmetry in knowledge – the machines knew a lot about our behaviour, but we knew very little about theirs. In a trade-off for convenience, these companies offered people information services in exchange for more information – data. The data has become more and more valuable, with Facebook making on average $30 from each of its 170 million American users. At the same time, we have fallen for the idea that these services are 'free'. In reality, we pay with our data into a business model of extracting human attention.

More data led to more profits, and so design patterns were implemented to encourage users to share more and more about themselves. Platforms started to mimic casinos, with innovations like the infinite scroll and addictive features aimed at the brain's reward systems. Services such as Gmail began trawling through our correspondence in a way that would land a traditional postal worker in prison. Live geo-tracking, once reserved for convicts' ankle bracelets, was added to our phones, and what would have been called wiretapping in years past became a standard feature of countless applications.

Soon we were sharing personal information without the slightest hesitation. This was encouraged, in part, by a new vocabulary. What were in effect privately owned surveillance networks became 'communities', the people these networks used for profit were 'users', and addictive design was promoted as 'user experience' or 'engagement'. People's identities began to be profiled from their 'data exhaust' or 'digital breadcrumbs'. For thousands of years, dominant economic models had focused on the extraction of natural resources and the conversion of these raw materials into commodities. Cotton was spun into fabric. Iron ore was smelted into steel. Forests were cut into timber. But with the advent of the internet, it became possible to create commodities out of our lives – our behaviour, our attention, our identity. People were processed into data. We would serve as the raw material of this new data-industrial complex.

One of the first people to spot the political potential of this new reality was Steve Bannon, the relatively unknown editor of right-wing website Breitbart News, which was founded to reframe American culture according to the nationalist vision of Andrew Breitbart. Bannon saw his mission as nothing short of cultural warfare, but when I first encountered him, Bannon knew that something was missing, that he didn't have the right weapons. Whereas field generals focused on artillery power and air dominance, Bannon needed to gain *cultural power* and *informational dominance* – a data-powered arsenal suited to conquer hearts and minds in this new battlespace. The newly formed Cambridge Analytica became that arsenal. Refining techniques from military psychological operations (PSYOPS), Cambridge Analytica propelled Steve Bannon's alt-right insurgency into its ascendancy. In this new war, the American voter became a target of confusion, manipulation and deception. Truth was replaced by alternative narratives and virtual realities.

Cambridge Analytica (CA) first piloted this new warfare in Africa and tropical islands around the world. The firm experimented with scaled online disinformation, fake news and mass profiling. It worked with Russian agents and employed hackers to break into opposition candidates' email accounts. Soon enough, having perfected its methods far from the attention of western media, CA shifted from instigating tribal conflict in Africa to instigating tribal conflict in America. Seemingly out of nowhere, an uprising erupted in America with manic cries of *MAGA!* and *Build the wall!* Presidential debates suddenly shifted from policy positions into bizarre arguments about what was *real news* and what was *fake news*. America is now living in the aftermath of the first scaled deployment of a psychological weapon of mass destruction.

As one of the creators of Cambridge Analytica, I share responsibility for what happened, and I know that I have a profound obligation to right the wrongs of my past. Like so many people in technology, I stupidly fell for the hubristic allure of Facebook's call to '*move fast and break things*'. I've never regretted something so much. I moved fast, I built things of immense power, and I never fully appreciated what I was breaking until it was too late.

AS I MADE MY WAY to the secure facility deep under the Capitol that day in the early summer of 2018, I felt numbed to what was happening around me. Republicans were already conducting opposition research on me. Facebook was using PR firms to smear its critics, and its lawyers had threatened to report me to the FBI for an unspecified cybercrime. The DOJ was now under the control of a Trump administration that was publicly ignoring long-held legal conventions. I had enraged so many interests that my lawyers were genuinely concerned the FBI might arrest me after I was finished. One of my lawyers told me the safest thing to do was stay in Europe.

I cannot, for security and legal reasons, quote directly from my testimony in Washington. But I can tell you that I walked into that room with two large binders, each containing several hundred pages of documents. The first binder contained emails, memos and documents showing the extent of Cambridge Analytica's data-harvesting operation. This material demonstrated that the company had recruited hackers, hired personnel with known links to Russian intelligence, and engaged in bribery, extortion and disinformation campaigns in elections around the world. There were confidential legal memos from lawyers warning Steve Bannon about Cambridge Analytica's violations of the Foreign Agents Registration Act, as well as a cache of documents describing how the firm exploited Facebook to access more than eighty-seven million private accounts and used that data in efforts to suppress the votes of African Americans.

The second binder was more sensitive. It contained hundreds of pages of emails, financial documents and transcripts of audio recordings and text messages that I had covertly procured in London earlier that year. These files had been sought by US intelligence and detailed the close relationships between the Russian embassy in London and both Trump associates and leading Brexit campaigners. This file showed that leading British alt-right figures met with the Russian embassy before and after they flew to meet the Trump campaign, and that at least three of them were receiving offers of preferential investment opportunities in Russian mining companies potentially worth millions. What became clear in these communications was how early the Russian government had identified the Anglo-American alt-right network, and that it may have groomed figures within it to become access agents to Donald Trump. It showed the connections among the major events of 2016: the rise of the alt-right, the surprise passage of Brexit, and the election of Trump.

Four hours went by. Five. I was deep into describing Facebook's role in – and culpability for – what had happened.

> Did the data used by Cambridge Analytica ever get into the hands of potential Russian agents? *Yes.*
>
> Do you believe there was a nexus of Russian state-sponsored activity in London during the 2016 presidential election and Brexit campaigns? *Yes.*
>
> Was there communication between Cambridge Analytica and WikiLeaks? *Yes.*

I finally saw glimmers of understanding coming into the committee members' eyes. Facebook is no longer just a company, I told them. It's a doorway into the minds of the American people, and Mark Zuckerberg left that door wide open for Cambridge Analytica, the Russians, and who knows how many others. Facebook is a monopoly, but its behaviour is more than a regulatory issue – it's a threat to national security. The concentration of power that Facebook enjoys is a danger to American democracy.

Dancing a delicate ballet among multiple jurisdictions, intelligence agencies, legislative hearings and police authorities, I have given more than two hundred hours of sworn testimony and handed over at least ten thousand pages of documents. I found myself travelling around the world, from Washington to Brussels, to help leaders unpack not only Cambridge Analytica but also the threats social media poses to the integrity of our elections.

Yet, in my many hours of giving testimony and evidence, I came to realise that the police, the legislators, the regulators and the media were all having a difficult time figuring out what to do with this information. Because the crimes happened online, rather than in any physical location, the police could not agree on who had jurisdiction. Because the story involved software and algorithms, many people threw up their hands in confusion. Once, when one of the law enforcement agencies I was dealing with called me in for questioning, I had to explain a fundamental computer science concept to agents who were supposedly specialists in technology crime. I scribbled a diagram on a piece of paper, and they confiscated it. Technically, it was evidence. But they joked that they needed it as a crib sheet to understand what they were investigating. *LOL, so funny, guys.*

We are socialised to place trust in our institutions – our government, our police, our schools, our regulators. It's as if we assume there's some guy with a secret team of experts sitting in an office with a plan, and if that plan doesn't work, don't worry, he's got a plan B and a plan C – someone in charge will take care of it. But in truth, that guy doesn't exist. If we choose to wait, nobody will come.

2

LESSONS IN FAILURE

It was eight years before this that I moved to England, where the story of my entanglement with Cambridge Analytica began. I'd worked in Canadian politics for a few years, but the irony is that I moved to London to escape politics. In the summer of 2010, I moved into a flat on the south bank of the River Thames, near the Tate Modern, the modern art museum housed in the colossal old Bankside Power Station. After several years in Ottawa, I had decided, at age twenty-one, to leave politics and move across the Atlantic to attend law school at the London School of Economics and Political Science (LSE). No longer in politics, I was unshackled from my old responsibilities to the party. It didn't matter anymore who I might be seen with, and I no longer had to watch what I said or think about who might be listening. I was free to meet new people and I was excited to make a new life for myself.

When I arrived it was still summer, and the first thing I did after unpacking was head out to sit with the sunbathers, tourists and young couples in Hyde Park. I took full advantage of London, spending Friday and Saturday nights in Shoreditch and Dalston, and Sundays in Borough Market, London's oldest food market, which is crammed into an outdoor hall alive with a cacophony of shouting traders, visitors and cooking stalls. I started making friends with people my age, and for the first time, I felt *young*.

But a few days after I arrived, still feeling fuzzy from the jet lag, I got a call that made it clear that it wouldn't be so easy to leave politics behind. Four months earlier, a man named Nick Clegg had become the country's deputy prime minister.

First elected to the European Parliament in 1999, he worked his way up to become, in 2007, the leader of the Liberal Democrats. This was back in the days when the Lib Dems were the radical third party of British politics – the first to support same-sex marriage, and the only party to oppose the war in Iraq and call for abandoning Britain's nuclear arsenal. In the 2010 general election, after more than a decade of Labour's now tired 'third way', 'Cleggmania' swept across Britain. At his peak, Clegg was polling as highly as Winston Churchill and positioned himself as Britain's answer to Barack Obama. After the election, he became part of a coalition government that made the Conservative David Cameron prime minister. The call was from his office: they'd heard about my data work in Canada and America from mutual connections in liberal politics, and they wanted to know more.

At the appointed time, I arrived at Liberal Democratic Headquarters (LDHQ), which was then still located at No 4 Cowley Street in Westminster. Located only a few blocks from the Palace of Westminster, the converted neo-Georgian mansion stood handsomely adorned in crimson brickwork and flanked by large stone chimneys on either side. It was rather outsize for the tiny, winding street, so I had no problem spotting it. Because it housed the offices of a party in Her Majesty's Government, an armed unit of the Metropolitan Police stood guard nearby, strolling up and down the small side street. After being buzzed in, I heaved open the weighty wooden doors and walked up to the reception, where I was greeted by an intern who would show me to the meeting. Still adorned with the manor's original chandeliers, oak panelling and fireplaces, the place revealed the faded elegance of a once grand residence, which felt oddly fitting for this once grand party.

Cowley Street, as they all called it, was unlike anything else I had seen in Canada or the United States. I wondered how all the party staffers waddling past one another in the cramped and creaking hallways could get anything done. Old bedrooms were jammed with desks, and cables connecting to servers were taped up on walls and around doorframes. In one converted closet, a man apparently with sleep apnea was snoring loudly on the floor, but no one paid him much attention. Looking around, I got the impression that this place operated more like an old boys' clubhouse than a party in government. I walked up a large staircase with ornately carved railings and was shown into a large boardroom that must have once been the main dining room. After I'd waited several minutes, a small cadre of staffers filed in. When the obligatory British small talk concluded, one of them said, 'So tell us about the Voter Activation Network.'

After Obama's 2008 victory, parties all over the world were becoming interested in this new 'American-style campaign', powered by national targeting databases and big digital operations. Behind the campaign was the emerging practice of *microtargeting,* where machine-learning algorithms ingest large amounts of voter data to divide the electorate into narrow segments and predict which *individual voters* are the best targets to persuade or turn out in an election. The Lib Dems wanted to talk to me because they were unsure whether they could translate this new school of campaigning into the British political system. What was so interesting for them about the project I had worked on with the LPC – setting up the same kind of voter-targeting system used by the Obama campaign – was that it was the first of its kind and scale outside the United States. And Canada, like Britain, uses the same first-past-the-post 'Westminster model' electoral system and has a diverse array of political parties. In this conversation, the staffers realised that half of the localisation work would have already been done if they imported the Canadian version of the technology. At the end of the meeting, they were almost giddy after learning about what the system could do. After leaving, I ran back to school to catch the tail end of a lecture on the rules of statutory interpretation and thought that was the end of it.

But the Lib Dem advisers called again the next day, asking if I could come back and tell my story to a bigger group. I was in the middle of a lecture, so I didn't pick up initially, but after four missed calls from a random number, I stepped out to see what was so urgent. There was a senior staff meeting that afternoon, and they asked if I could do an impromptu presentation on microtargeting. So after class, I walked from LSE back to Cowley Street with my backpack filled with textbooks. With such short notice, I didn't have time to change, so I headed to meet the deputy prime minister's advisers in a Stüssy graphic print T-shirt and camo sweatpants.

Walking into the same boardroom, I was met this time with the humming cacophony of a packed room. I was ushered to the front without any prompt, so, after apologising for my slightly ridiculous attire, I proceeded to just wing it. I told them how the Lib Dems could use microtargeting to overcome the disadvantages that come with being such a small party. And as I continued, I couldn't help but get more passionate. I hadn't spoken about this since I left the LPC, and my heart simply poured out. I told them about what I had seen on the Obama campaign, what it was like to see so many people vote for the first time, what it was like to see African Americans at rallies filled with *hope*. I told them that this was not just about data; this was about how we could reach the people who had given up on politics. This was how we would find them and inspire them to turn out. But, most important, this was about how technology could be the vehicle that this party, which now found itself in the corridors of power, deployed to upend the entrenched class system that underpins so much of British politics.

A few weeks later, the Lib Dems asked me to come work for them and implement a voter-targeting project in Britain. I was just starting my degree at LSE, and, as a twenty-one-year-old student, I was finally finding my feet in London. I hesitated about whether it was really a good idea to distract myself with politics again. But here was a chance to take the same technology – the same software, and essentially the same project – and finish what I had started in Canada. But it was what I saw hanging so casually on the wall in one of the offices at Cowley Street that finally pulled me in. It was an old yellowing card, with slightly curled corners, with an excerpt of the Liberal Democrats' constitution that read NO ONE SHALL BE ENSLAVED BY POVERTY, IGNORANCE OR CONFORMITY.

I said yes.

AFTER THE 2008 PRESIDENTIAL ELECTION, I returned to Ottawa and wrote a report about the Obama campaign's new technology strategies. It landed with a thud. Everyone was expecting me to tell them about the campaign's flashy branding, graphics and viral videos.

Instead I wrote about relational databases, machine-learning algorithms, and how these things connected to one another through software and fundraising systems. When I recommended that the party invest in *databases,* people thought I'd lost my mind. They wanted sexy answers – not *this*. Obama was their benchmark for a 'model' campaign, and they were taken with high cheekbones and pouty lips, not the skeleton and backbone that made it all possible.

Most campaigns can be boiled down to two core operations: *persuasion* and *turnout*. The turnout, or 'GOTV' (get out the vote), universe is those people who likely support the candidate but do not always vote. The persuasion universe is the inverse, representing those who likely will vote but do not always support the party. People who are either very unlikely to vote or very unlikely to ever support us are put into an exclusion universe, as there is no point in engaging them. Voters who are both very likely to support the candidate *and* very likely to vote – these are the 'base' voters, and they are typically excluded from contact, but they might be prioritised for volunteer or donor recruitment. Finding the right set of voters to contact is the name of the game.

In the 1990s American voters were generally targeted using data provided by local or state offices, which typically contained each voter's party registration (if they had one) and their voting history (which elections they came out to vote in). However, the limitation to this approach is that not all states provide this information, voters change their mind more frequently than they change their party registration (or would not register a party), and this information would tell you nothing about the issues that actually motivate the voter. What microtargeting did was find extra data sets, such as commercial data about a voter's mortgage, subscriptions, or car model, to provide more context to each voter. Using this data, along with polling and the statistical techniques, it's possible to 'score' all of the voter records, yielding far more accurate information.

What Obama's campaign did was mainstream this technique and put it at the heart of its campaign operations. This is important because the organised chaos of campaign activity is typically not what one sees on TV, such as speeches or rallies. Rather, it is the millions upon millions of direct contacts made by volunteer canvassers or via direct mail to individual voters throughout the country. Although less sexy than a beautifully crafted speech or amazing branding, it is this unseen machinery that provides the critical horsepower of a modern presidential campaign. When everyone else is focused on the public persona of the campaign, strategists are focused on deploying and scaling this hidden machinery.

Eventually some of us in the Opposition Leader's Office, where I was working in Parliament, realised we could show the party how useful the Voter Activation Network would be if we created a parliamentary version of it for the leader's interactions with constituents and citizens. The party was unwilling to foot the bill for something so extravagant as a new database, but we realised we had room for it in the leader's official parliamentary budget. The only problem was that these were technically public funds, and any pilot database we created could not be used for political purposes. But we were not too concerned. A parliamentary version would contain the records of constituents and citizens who had contacted the leader, and since constituents are simply voters with a different hat on, it would allow us to highlight all the same functionalities to the party without needing them to spend anything. Surely, after seeing such a system firsthand, the Liberal Party of Canada would begin to understand the potential of data. We asked Mark Sullivan and Jim St George if they'd ever thought about expanding VAN internationally – to Canada. They hadn't done any big projects outside the United States at that point, but they jumped at the chance to work with us. With the help of Sullivan and St George, we were able to create a Canadianised VAN infrastructure in six months. To the party's delight, VAN even worked in both English and French. There was only one problem: *there was no data to actually fuel the system.*

Computer models are not magical incantations that can predict the world – they can make predictions only when there is an ample amount of data to base a prediction upon. If there was no data in the system, then there could be no models or targeting. It would be like buying a race car but skimping on gasoline – no matter the car's sophisticated engineering, it just wouldn't start. So the next step was to procure data for VAN. But data was going to cost money, and because it would be used for campaigns, by law the party had to pay for it, not the leader's parliamentary office. But almost immediately there was hard pushback from the party, which was not eager for change. I turned to the MP who first brought me into politics, Keith Martin. He had given me my first internship when I was still in school, and later my first real job, in the Canadian Parliament. Martin was often called the 'maverick' of Canadian politics, and he staffed his office with mavericks, too. For me, he was a perfect fit. Martin trained as an ER doctor and spent his early medical career in African conflict zones, treating everything from land-mine injuries to malnutrition. This guy was legit cool and lived an amazing life before politics – on the wall in his office he had photos of himself looking like Indiana Jones in a khaki overshirt, sitting with leopards. As an ER medic, he was trained to not waste time, but in politics you survive by wasting time. He was once so incensed at the mechanistic procedures of Parliament that, mid-debate, he picked up 'the Mace' – the gold-plated medieval weapon we inherited from Britain that lies in the aisle of the House of Commons.

In 2009, Jeff Silvester, Martin's senior adviser, a former software engineer who'd turned his attention to politics, was one of the few people in the party who understood what I was trying to do. He was my mentor and my rock throughout my time at Parliament. I explained that even though the party hadn't authorised me to move forward with the data targeting program, we needed to do it. And that meant we needed funding. With Martin's approval, Jeff agreed to help me raise money without telling the party's national office about it. We started holding secret events where I could explain to would-be donors that we needed this program if the LPC had any hope of being competitive in the twenty-first century. We did all this on the down-low, persuading people on the ground to put in money while the party staff weren't paying us any attention. In short order, we raised several hundred thousand Canadian dollars, which was enough to get the program started. Dissatisfied with the national office, the British Columbia wing of the party agreed to be the guinea pigs in our experiment.

It was not clear whether any of this would work. In the United States, there are only two major parties, whereas in Canada there are five. This means that the dimensionality of what you are predicting for is no longer binary (Democrat or Republican) but multivariate (Liberal, Conservative, New Democratic Party, Green or Bloc Québécois). With more options, you get many different kinds of swing voters (e.g., Lib-Con *vs* Lib-NDP *vs* Lib-Green, etc.) who can float their support around in many directions. There was also a far less developed market for consumer data in Canada and Europe, so many of the standard data sets in the USA either were not available or had to be pieced together from many sources. Finally, parties in other countries often have strict donor or spending caps. A lot of people were sceptical that microtargeting could even be deployed outside of America, but I wanted to try nonetheless.

I called up Ken Strasma, who ran Obama's targeting operation in 2008, and asked if he'd be willing to help us set up a program in Canada. Strasma's team in Washington, DC, then built the models. The BC office in Vancouver pieced together useful data sets, such as old polling and canvassing data, and Strasma worked out how to deal with the additional complexities of multi-party politics. Volunteers all over the province were given either these new canvass lists to try out or old lists that would serve as controls. A sigh of relief came to the BC party staff as the results came in. As with many campaigns, the party uses persuasion canvasses to reach out to voters who have not yet made up their minds which party they will be supporting. By comparing the successful conversion rates (where a previously undecided voter declares support for your party) of the old lists with those of the new microtargeting lists, the BC operation was able to establish that the new targeting approaches had higher conversion rates. It was very exciting. We proved that what Obama accomplished in America would be possible in different political systems around the world. But when the national party in Ottawa found out what we had done, they objected to a national project. They wanted to run campaigns like Obama's, but when shown how to do it, they refused.

I'd been drawn to politics because it seemed like a way to make a difference in the world, but after more than a year of banging my head against the wall, what was the point? Then came an intervention. In the Liberal Party, many of the secretarial staff came from a gaggle of older Québécois ladies, who had been around long enough to see how politics can transform someone. They took me to lunch in Gatineau, the French part of town just across the Ottawa River. After lighting up their cigarettes, they said, in their raspy, accented voices, *'Listen, don't become like us.'* They told me they'd given their lives to the party, that the party had given them nothing in return except for 'expanded waistlines and several divorces'. *'Go, be young'*, they said. *'Get the hell out of here before it traps you.'* And I realised they were right. At just twenty years old, I was already a midlife crisis waiting to happen.

I decided on law school at the London School of Economics, as London was probably far enough from Ottawa – 3,300 miles and five time zones across the Atlantic. I later learned that some of Canada's party leadership had conflicted motives. The party still awarded many of its advertising, consulting and printing contracts to firms owned and operated by senior party members or their friends. A new, data-centric approach might mean that 'friends and family' in the party would lose out. In 2011, one year after I left Ottawa, the LPC was devastated in the federal election by the Conservative Party of Canada, which had invested in sophisticated data systems at the behest of its imported Republican advisers. For the first time ever, the Liberal Party was relegated to third place, with only thirty-four seats in Parliament. It was an historic defeat.

WHEN I BEGAN WORKING for the Liberal Democrats in London, it was for only a few hours a week, in between my classes at LSE. But almost right away I realised that, compared with the Obama campaign or even the LPC, the Lib Dems were a complete train wreck of a party. The office operated more like a stale curiosity shop than the heart of a political machine. The LDHQ staff were mostly bearded men in suits and sandals who spent more time chitchatting about the old Whigs than doing anything to mobilise their campaign. I asked to see their data systems, and someone told me about EARS, which stood for Electoral Agents Record System. 'Wow, okay. This looks ... old-school,' I said. 'Was this made in the eighties?' It was like asking for a graphics demo and being shown one of those old *Pong* games. Someone told me that one of the systems had been designed during the Vietnam War.

It soon became clear to many in the party how superior VAN was to anything else available, and the party finally approved a contract with VAN to set up the data infrastructure. But now we needed data – the fuel to run the Ferrari. This was the step where the project had gone wrong in Canada, and the process went no more smoothly in the UK. There is no national electoral register in the UK – it's all handled by town councils – so we had to approach hundreds of different councils all across Britain to get their voter data. I'd be on the phone to Agnes in West Somerset, who would be like 105 years old and had probably been managing the voter rolls since women got the vote, asking her, 'Do you have a digital copy of the register?' No, she'd say, because she kept the records as they'd always been kept, on paper, but I could see a copy in a bound book in the local town hall. Sometimes the local officials agreed to give us the data, sometimes not. Sometimes it was in electronic form, sometimes it was a PDF file, and sometimes it was just reams of paper that we had to feed into an optical scanner. Excel files would usually be emailed without a password – because why would anyone want to steal voter data?

The British electoral system was stuck in the 1850s, and, as I soon found out, so were the Lib Dems' tactics. It wasn't hard to understand why the party and its old Liberal Party predecessor had been on a losing streak since World War II. Leaders had lost touch with how to win and were utterly obsessed with handing out leaflets. These leaflets were called 'Focus' and usually complained about parochial 'local issues' like potholes or rubbish collection. The Lib Dems thought this was a clever way of 'slipping in' their messaging to something that looked like a local newspaper. But there was a problem with the Lib Dems' dollar-store *Pravda*: no one actually read it. Their idea of a voter was someone who spent their weekends flipping through mail order catalogs and political literature – political staffers are often so socially clueless, they forget that regular people have lives. Despite being the smallest of the three main parties, the Lib Dems had the most volunteers, because they were absolutely militant about shoving leaflets through doors, rain or shine. They would decide how many leaflets to deliver before even deciding what they should say in them.

In the world of hacking, the term 'brute force' refers to randomly trying every possible option until you hit on the correct one. It doesn't involve strategising – it's simply throwing everything at the wall to see what sticks. That's essentially what the Lib Dems were doing, spending tons of money on leaflets without targeting particular voters. Brute force is an unsophisticated hack, comically inefficient yet occasionally successful. There were certainly more effective ways to win elections. Yet when I tried to present alternatives to spamming voters with their upper-middle-class propaganda, complete with MS Word clip art, I would get a lecture on 'how Lib Dems win' and the fabled Eastbourne by-election, in 1990 – the surprise victory of the first MP to win as a Lib Dem since the 1988 SDP-Liberal merger – where they had apparently delivered *a lot* of leaflets. To question Eastbourne was heresy. Fringe religions demand conformity, and the Lib Dems were no different. The party was a leafleting cult.

By-elections are irregular special elections, such as a vote to replace a member of Parliament who's died. The Lib Dems were obsessed with them. For some reason, whenever the party won a by-election, members acted as if they'd conquered Britain with banners that proclaimed LIB DEMS WINNING HERE! But as part of my research, I decided to catalog every election and by-election since 1990 and found that the party had lost the overwhelming majority of them. 'But what you are doing doesn't help you win. It helps you lose. Here is the data,' I told them. 'These are facts.'

Some of the party leadership listened, but most just seemed pissed off. They had their own cottage industry of being 'election gurus' for the party faithful and weren't keen on a newcomer strolling in and telling them how to fix their system. This did not bode well.

In the meantime, I started playing around with the voter data we had been able to collect from data vendors like Experian. I experimented with different types of sandbox models, similar to what I'd done in Canada. And something strange kept happening. No matter how I designed the models, I couldn't build one that reliably predicted Lib Dem voters. I had no trouble doing it for Tories or Labour. Posh dude in some leafy rural suburb? *Tory.* Live on a council estate in Manchester? *Labour.* But Lib Dems were these in-between weirdos who resisted any neat description. Some of them looked Labour-*ish* and some of them looked Tory-*ish*. *Ugh, what am I missing?* I wondered if there was a latent variable at work. In social sciences, a 'latent variable' is an element that is influencing a result, but one you haven't yet observed or measured – a hidden construct that's floating just out of view. *So what is the hidden construct here?*

One problem was that, at a basic level, I couldn't visualise a Lib Dem voter. I could visualise Tories, who – in the most general sense – were either posh, rich, *Downton Abbey* types or working-class, anti-immigrant types. Labour voters were northerners, union members, council estate dwellers, or public-sector types. But who were the Lib Dems? I couldn't imagine a path to victory if I couldn't imagine who'd be marching with us on that path.

So, in the late spring of 2011, I started travelling around Britain to find out. For several months, I'd go to my classes at LSE in the morning, and then in the afternoon I'd hop on a train to places with delightful names like Scunthorpe and West Bromwich and Stow-on-the-Wold. My intent was to do voter interviews and focus groups, but not the usual kind. Instead of asking prepared, scripted questions, I'd have unstructured conversations, so people could tell me about their lives and what was important to them. I could have jumped directly to polling, but I realised that any questions I asked would be biased by what I, the questioner, thought was the most relevant thing to ask. Sure, I would get answers to the questions I added to the poll, but what if they were the wrong questions? I went to speak to people because I knew I was already biased and coloured by my own experiences. I did not know what life was really like for an older British man living on a council estate in Newcastle, or a single mother of three in Bletchley. I wanted them to tell me what they wanted me to know about their lives, in their own words and on their own terms. So I got local constituency parties and polling firms to help randomly select people to speak to.

For the focus groups in smaller villages, there were often no addresses. I'd show up and be told, 'We're meeting in the cottage up on the hill. Just walk past the pub, through those fields with the daffodils, and after a while you'll see it.' Random townspeople would show up, and perhaps Clive the local barman or Lord Hillingham the gentleman farmer would amble by. Sometimes I would just go to the village pub and chat with people there. British people were whimsical, nuanced, and often fun to chat with, and the focus groups reminded me of the town halls I had loved so much back in BC. People would talk, and I'd just listen and take notes on what they had to say.

Through these many conversations I travelled alone, as the party was not terribly interested in what I was up to, but I started to piece together the randomness of the Liberal Democrats. What quickly became apparent was that they lived so many different lives. They were farmers in Norfolk in tartan hats. Hipsters being artsy in Shoreditch. Old Welsh ladies in the Mumbles or Llanfihangel-y-Creuddyn. Gays in Soho. Professors at Cambridge who hadn't brushed their hair in twelve years. Lib Dem voters were an odd, eclectic mix.

They may have all looked different, but I noticed that they did have one common trait. Labour Party voters would say, 'I'm Labour.' And the Conservatives would say, 'I'm Tory.' But the Liberal Democrats would almost never say, 'I'm Lib Dem.' Instead they would say, 'I *vote* Lib Dem.' This was a slight but ultimately important distinction. It took me some thinking to figure out that it might have to do with the party's history. The party wasn't officially formed, in its modern incarnation, until 1988, after the merger of two smaller parties, which meant that many of its current voters originally came from historically Tory or Labour families. This meant that at some point in their lives, they'd had to make an active decision to switch from an old party to this new party. For them, supporting the Lib Dems was an *act*, not an *identity*.

ONE OF THE PEOPLE who drew me to London was Mark Gettleson, who quickly became one of my best friends. I had first met Gettleson in, of all places, Texas. Back in 2007, when I was just starting out at the LPC, they sent me to a Democratic Party event in Dallas to do some networking. I was milling about with hundreds of people in a giant ballroom, marvelling at all the Stetson hats, when a clipped British voice behind me said, '*Youuuu* are not from here.' I turned around to see a bloke grinning like the Cheshire cat in forest-green trousers and a floral Liberty-print shirt. Between my bleached platinum-blond hair, complete with a classic mid-noughties fringe, and his dandified manner of dress, we were drawn together like two butterflies at a moth convention.

The son of a family of Jewish antique dealers on London's Portobello Road, Gettleson is posh, eccentric and delightfully camp, and speaks with a delivery reminiscent of the actor Stephen Fry. In another time, he would have been a dandy in the salons of eighteenth-century London. He's a polymath of the highest order. In conversation, he can draw connections between early-1990s hip-hop and the Franco-Prussian War without taking a breath. Gettleson and I vibed that night, and over the next couple of years I'd see him at various political gatherings in America or Britain. After I decided to move to the UK, we immediately started to hang out, sometimes in the converted crypt underneath an old church that he'd somehow turned into a fabulous flat with bizarre antique miniatures and art lying around everywhere in a chaos that somehow worked. 'I don't do minimalism, Chris. *I'm a maximalist,*' he would say as I looked through all his objects.

I FLOURISHED IN LONDON and quickly gained a wide circle of friends. Although I was studying law at LSE and working at Parliament, most of my friends were a gaggle of club kids, dancers, queens, flamboyant creatives and design students from Central Saint Martins, one of the world's preeminent design and fashion schools, which had produced graduates like Alexander McQueen, John Galliano and Stella McCartney. The thing that was different about Gettleson was that, like me, he seamlessly floated between all these different worlds. At the time, he was working for the London office of Penn, Schoen and Berland, the well-known Democratic polling firm that had most famously once been affiliated with the Clintons. He was the *only* person I knew who could join me for a formal reception on the Terrace Pavilion at Parliament, surrounded by cabinet ministers, and then end up with me later in the night decked out in makeup, glitter and wigs, voguing among a heaving cavalcade of queens at a Sink the Pink glam ball. Gettleson was magnetic and all my friends adored him. As gentle as he was exuberant, he shepherded twinks on nights out like a collie shepherds lambs. They would find themselves completely mesmerised by his spontaneous use of Barbie dolls and character voices to explain why Neville Chamberlain's policy of appeasement failed – at 4 a.m. at the height of a thumping house party.

Gettleson was also one of the few people who understood what I was trying to do with data. One afternoon, as I was complaining about my difficulties in constructing a model for Lib Dems' voting behaviour, I told him I was thinking of asking some Cambridge profs about it. He connected me with Brent Clickard, who was completing his PhD in experimental psychology at Cambridge and might be able to introduce me to some professors there. Clickard turned out to be so much more than simply a conduit to Cambridge. Like Gettleson, he was a dandy, the kind of guy who dresses in tweed and always has a crisp paisley pocket square. Though he came from a wealthy Midwestern American family, he spoke with a delightfully affected mid-Atlantic accent that he somehow picked up, as if he were playing a character in *Casablanca*. He'd been a dancer with the Los Angeles Ballet before deciding to move to England.

In the course of several boozy conversations, Clickard suggested I look more deeply into personality as a factor in voting behaviour. Specifically, he pointed me to the five-factor model of personality, which represents personality as a set of ratings on five scales: openness, conscientiousness, extraversion, agreeableness and neuroticism. With time and testing, the measurement of these five traits has proven to be a powerful predictor of many aspects of people's lives. A person scoring high in conscientiousness, for example, is more likely to do well in school. A person scoring higher in neuroticism is more likely to develop depression. Artists and creative people tend to score high in openness. Those who are less open and more conscientious tend to be Republicans. This sounds simple, but the Big Five model can be an immensely useful tool in predicting voters' behaviour. In political discourse, you find that many of the phrases used to describe candidates, policies, or parties align with personality. Obama ran on *change, hope* and *progress* – in other words, a platform of openness to new ideas. Republicans, on the other hand, tend to focus on *stability, independence* and *tradition* – in effect, a platform of conscientiousness.

Reading in my flat in the middle of the night, I finally realised something. Maybe the Lib Dems didn't have a geographic or demographic base; maybe they were a product of a *psychological base*. I put together a pilot study and found that Lib Dems tended to score higher on 'openness' and lower on 'agreeableness' than Labour or Tory voters. I realised that these Lib Dems tended to be, like me, open, curious, eccentric, stubborn and a bit bitchy at times. This is how an artist in East London, a professor at Cambridge, and a farmer in Norfolk could all coalesce around this party in their own way, despite living very different lives.

The five-factor model was the key that cracked the Lib Dems code – and, in the end, provided the central idea behind what became Cambridge Analytica. The five-factor model helped me understand people in a new way. Pollsters often talk about monolithic groups of voters – women voters, working-class voters, gay voters. Although certainly important factors to people's identities and experiences, there is no such thing as a *woman voter* or a *Latino voter* or any of these other labels. Think about it: if you randomly grab a hundred women off the street, will they all be the same person? What about a hundred African Americans? Are they all the same? Can we really say that these people are clones by virtue of their skin colour and vaginas? They all have different experiences, struggles and dreams.

Exploring the nuances of identity and personality started to help me unpack why, despite the fact that politicians do polling all the time, they still seem horrendously out of touch. This is because so many of their pollsters are out of touch. Polling firms influence politicians' ideas of what makes up voter identity, which are usually horrendously oversimplified or just plain wrong. Identity isn't ever a single thing; it's made up of many different facets. Most people do not ever think of themselves as a *'voter'*, let alone curate an identity around how their worldview relates to tax policy. When a person goes grocery shopping, they are unlikely to stop, drop their shopping in a moment of blinding self-awareness, and suddenly realise in the middle of the store that they are, in fact, a *university-educated white suburban female in a swing state*. Whenever I was doing focus groups, people tended to talk about how they grew up, what they do, their families, what music they like, their pet peeves and their personality – the kinds of things you talk about on a first date. Can you imagine how terrible a blind date would go if you were allowed to ask only the standard polling questions? Yeah, exactly.

IN LATE 2011, I broke the news to Nick Clegg's team that I thought the party was in deep trouble. I explained that the data showed that Lib Dem voters were ideological, they were stubborn, and they hated compromise. But the party had become the antithesis of these attributes when it joined a coalition government with the Tories. The party was composed of uncompromising supporters, and yet it was operating in a government birthed out of compromising its principles. This type of compromise was a betrayal of Lib Dem voters' ideals, and it was bound to drive people out of the party.

I put together slides and gave a presentation to Lib Dem leaders in an old wood-panelled committee room at Parliament. They'd been called together to hear an interim update on what I was finding and were excited to hear about what all this new technology was discovering. But their smiles quickly evaporated, as it was all very doom-and-gloom, describing in detail the tactical deficiencies in the party's strategy. I created one slide showing that both Labour and the Tories had extensive data coverage of the voting population, meaning that they had quite a bit of data recorded for each voter, whereas the Lib Dems covered less than 2 per cent of it. The report was damning and embarrassing, and no one wanted anything to do with it – or, ultimately, with me. It's fair to note here that I can be a bit blunt at times and have a tendency to piss people off. I'm a bit like Marmite, people either love it or hate it, but no one is ever blasé about Marmite. Suffice it to say that the party stalwarts were not keen on having some random Canadian who looks like an intern sashaying in and telling them they're doing everything wrong.

The one Lib Dem who listened was the chief whip, Alistair Carmichael. Carmichael is as Scottish as they come, hailing from Islay, the southernmost island of the Inner Hebrides. A native Scot who grew up speaking Gaelic in school, Carmichael talks with a Highlander brogue mixed with a more 'proper' Edinburgh accent he picked up in his early years as a crown prosecutor. He is chatty and warm, and when I visited his office he always invited me to join him in a tipple of whisky from his well-stocked cabinet. As a government whip, he was a hardened political machinator whose easy manner belied a profound understanding of the levers of power. His position as chief whip meant he'd seen and heard everything, so I turned to him for advice on how to move beyond the impasse I was experiencing in the party. I always felt like I could speak frankly to Carmichael, which he, as a man who did not have a fear of speaking his mind, respected. And he tried, unfortunately to little avail, to persuade the party staff to heed what I was saying.

All of this was beyond frustrating. I was showing them data, supplemented by peer-reviewed literature. I was showing them *science*. And they were responding by calling me pessimistic, problematic, not a team player. The last straw came when someone leaked my slides, apparently in an attempt to embarrass me. It backfired when a journalist wrote approvingly of my arguments, noting that the Lib Dems suffered from 'the great leafletting problem' and were far behind the Tories and Labour in data collection and research. When you spend so much time researching voters and going out and meeting them, you grow more and more connected to them. I felt like my work was not just about winning an election; it was also about understanding what the lives of people were really like. It was about expressing and reiterating to those in power what it was like to be trapped by poverty, ignorance, or conformity.

Two years later, in 2014, the Liberal Democrats lost 310 council seats and all but one of their twelve seats in the European Parliament. The coup de grâce then came in May 2015, when the party was eviscerated, losing forty-nine of its fifty-seven seats in Parliament. With only eight Lib Dem MPs being reelected, their entire parliamentary caucus could have comfortably fit into a Mazda Bongo camper van.

3

WE FIGHT TERROR IN PRADA

LONDON'S MAYFAIR NEIGHBOURHOOD IS A PLACE OF EXCEPTIONAL wealth and power, with an unabashed legacy of empire. Walking down its old streets, one can see dozens of blue circular plaques dotting the buildings, commemorating the famous playwrights, authors, politicians and architects who once inhabited this place. Sitting in Mayfair's southeastern corner, not far from 10 Downing Street, is St James's Square, which is lined with grand old Georgian row houses. On the square's north end sits Chatham House, the location of the Royal Institute of International Affairs. On the east end sits the headquarters of British Petroleum, or BP, one of the largest oil companies in the world, and Norfolk House, which during World War II served as the offices for US general Dwight D. Eisenhower and the Supreme Headquarters Allied Expeditionary Force. Several private clubs dot the square, including the colonial-era East India Club and the Army and Navy Club. At the centre of the square is a small garden encircled by sections of decorative iron fencing. And standing in the middle of the park is an equestrian statue of William III looking out toward the buildings. The centre garden is surrounded by shrubbery and tidily groomed flower beds. St James's Square is a living monument to the global dominance of British colonialism.

South of BP, on the east end of the square, there is a building, several stories high, that dates to 1770. It's built of smooth grey sandstone, with flaxen brick and a pair of stone Ionic pillars that flank the entrance. A red Royal Mail postbox stands outside. This, in early 2013, was the headquarters of the SCL Group. Originally known as Strategic Communication Laboratories, the firm was led by Nigel Oakes and had existed in various forms since 1990. SCL was cleared by the British government to access 'Secret'-level information, and its board included Thatcher-era former cabinet ministers and retired military commanders, as well as professors and foreign politicians. The firm worked primarily for militaries, conducting psychological and influence operations around the world, such as jihadist recruitment mitigation in Pakistan, combatant disarmament and demobilisation in South Sudan, and counternarcotics and counter-human trafficking operations in Latin America.

I heard about them in the spring of 2013, a few months after I left the Liberal Democrats, when a party adviser I kept in touch with called to tell me about an opening. He said he thought of me because this firm was looking for 'data people for some behaviour research project' involving the military. It never occurred to me to work on defence projects, but after two failures with political parties, in both Canada and Britain, I was ready to try something new.

Walking in through the front door, I entered a lobby with checked black-and-white marble floors, a crystal chandelier and ornate plaster fringes on the cream-coloured walls. The offices have kept many of the original details of the building, with several rooms centred around a marble fireplace. Tightly woven green carpets with tiny red-and-white circular frills lined the floors. I was then shown to a small room where I was told to wait for a man named Alexander Nix, who was one of the directors of the SCL Group. I remember it being exceptionally hot, as if the heating had been left on high even though it was late spring. (I learned later that the heat was intentional – a way to mess with people before a meeting.) I sat in that tiny sweatbox of a room for about ten minutes, until a man entered. The first thing I noticed was his impeccably tailored Savile Row suit, over a shirt embroidered with his initials. His eyes were sapphire, in striking contrast to his pale, paperlike skin.

It was the perfect setting for my introduction to Alexander Nix, who was born into the British upper classes and schooled at Eton, where the royals have sent their children and whose uniform still includes collars and tails. Most English aristocrats have an air of camp to them, and in this tradition, Nix did not disappoint. His accent was as rich as they come. He wore black-rimmed glasses, and his floppy strawberry-blond hair had a deliberately casual little flip to it. He invited me to sit down amid piles of paper and boxes – detritus from a recently finished project, he told me. They planned to move soon to a larger office.

It didn't take long before Nix was describing the ins and outs of SCL's business. He got me to sign a non-disclosure agreement (NDA) and proceeded to tell me that most of the firm's work was for military and intelligence agencies, on projects that governments couldn't officially undertake themselves. 'We win hearts and minds over there … *you know,* however that needs to happen.' He pointed to a framed photo of a rally in what appeared to be a country somewhere in Africa.

When I asked for details, he produced a few reports. As I flipped through them, he started to explain TAA, or target audience analysis. This is the first step in an information operations project, he said – analysis and segmentation. But as I skimmed the reports, I couldn't believe how crude the methodology was – and I wasn't shy about saying so.

'This could be done so much better,' I told him. As I'd soon learn, Nix could lose his temper in the moment it took to register a challenge to his inborn sense of superiority. For now, he merely bristled.

'We,' he said, 'are the best firm at doing this.'

'Sure,' I said, 'but you could be doing a lot better job of targeting. It looks like the army is literally dumping leaflets out of a plane. If the army has laser-guided missiles, why are you doing this with propaganda?' It was a harsh response – especially in a potential job interview – and Nix was taken aback. *I'm the one who does the talking,* he seemed to be thinking. The conversation ended abruptly, and as I headed out, I thought, *What a massive waste of time.*

But it wasn't. Nix called soon after to ask if I'd be willing to talk some more, to explain what I thought SCL was doing wrong and how they might fix it.

THE WORLD OF PSYCHOLOGICAL warfare of which SCL was a part has been around for as long as humans have waged war. In the sixth century BC, Persians of the Achaemenid, knowing that Egyptians worshipped the cat god Bastet, drew images of cats on their shields so the Egyptians would be reluctant to take aim at them in battle. Rather than simply destroy and pillage enemy cities, Alexander the Great used positive psychological tactics, leaving troops behind to spread Greek culture and assimilate the defeated into his vast empire. During the Middle Ages, Tamerlane and Genghis Khan used terror as a psychological weapon, decapitating foes and parading their severed heads around on pikes. And in Russia, Ivan the Terrible cowed the masses into submission by setting up giant frying pans on Red Square and roasting his enemies alive. During World War II, the British perfected the art of misdirecting the enemy by staging fake invasions, using dummy tanks, and even planting fake battle plans on a corpse dressed as a dead soldier in the fantastically named Operation Mincemeat. The well-designed use of information – and disinformation – is one of the most effective ways of gaining tactical advantage on the battlefield.

In devising an informational weapon, it's helpful to think of the basic aspects of any weapon system: the payload, delivery system and targeting system. For a missile, the payload is an explosive, the delivery system is a rocket-propelled fuselage, and the targeting system is a satellite or a heat-seeking laser. With informational weapons, the same components are present. But there is one key difference: the force you are using is *non-kinetic*. In other words, you don't blow stuff up. In informational combat, the payload is often a story – a rumour deployed to trick a general or a cultural narrative intended to pacify a village. And just as the military invests in chemistry to inform bomb building, it also tries to research what kinds of narratives will yield the biggest impact.

Historically, US leaders have undervalued information operations, thanks to a robust advantage in missiles, tanks, bombers, ships and guns. The United States has undertaken some information operations, though mostly of the age-old paper-leafletting variety. In the Korean War, US troops used loudspeakers to blare propaganda, while aircraft scattered leaflets across enemy lines. During the Vietnam War, specialised PSYOPS battalions planned similar propaganda blitzes, aiming to win as many 'hearts and minds' as possible. But bolstered by an unmatched defence budget, the American military has become a gang of boys with toys, where force amplification is physical and kinetic.

Tanks and bunker busters are useless against viral propaganda and web-fuelled radicalisation. ISIS doesn't just launch missiles; it also launches narratives. Russia compensates for its aging military equipment with 'hybrid approaches' of attack, beginning with the ideological manipulation of target populations. Terror groups use social media to recruit new members, who then use guns and bombs to achieve their ends. These threats are no less dangerous for being nonconventional, and western powers have struggled to respond. You can't shoot a missile at the internet, and traditional American military culture, dominated by straight white men who like to give and take orders, is hostile to the kind of nonconformist recruit who might introduce more nuanced, tech-enhanced counterattacks.

The US military's Defense Advanced Research Projects Agency (DARPA) has tried to grapple with these new realities of terror and conflict. Among the stated goals of past DARPA-run programs – with names like *Narrative Networks* and *Social Media in Strategic Communications* – has been to 'attempt to track ideas and concepts to analyse patterns and cultural narratives' and 'develop quantitative analytic tools to study narratives and their effects on human behaviour in security contexts.' The US military also ran a program called the *Human Social Culture Behaviour Modelling Program,* which aimed to create 'tools for sociocultural analysis and forecasting to users in the field.' In other words, the point of many of these programs is to gain *total informational asymmetry* against threats – to have so much data that we would be able to completely overwhelm and dominate the information space surrounding our targets. This was the lucrative niche Nix had his eye on to win new contracts for SCL.

NIX INITIALLY OFFERED ME a three-month contract to do, essentially, whatever I wanted. 'I'm not even going to create a job description,' he said. 'Because, frankly, I don't know what I would put in it.' After all the agonies of dealing with the LPC and the Lib Dems, it was incredibly enticing to be given free rein. So in June 2013, I started working with SCL.

Like most people, I had never paid much attention to military strategy, save for the occasional midnight binge watch of the History Channel. With such a daunting learning curve, I needed to quickly get up to speed on the firm's current projects. The problem? No one would answer my questions. In fact, my new colleagues couldn't close their laptops fast enough. 'Why do you need to know?' they'd say, or 'I need to check on whether I can talk to you about that.' The secrecy was going to make it difficult to figure out how to do whatever I was supposed to be doing. When I complained about this to Nix, he elaborately rolled his eyes and simply handed me the keys to a cabinet in his office. Inside I found binders of old reports.

The documents described projects SCL had undertaken for its old clients, which included the British Ministry of Defence and the US government. It was working in Eastern Europe for NATO on counter–Russian propaganda initiatives. One was for a counter-narcotics program in a Latin American country, where a military client spread disinformation to turn local coca farmers against drug lords. Others detailed PSYOPS programs in Mexico and Kenya. As Nix had said before, these were projects that government agencies did not want to officially undertake themselves. Rather, they would hire contractors to enter the region as a 'market research firm' or under some kind of faux-business cover.

One report that caught my eye described a Ministry of Defence project for using information operations to influence different target groups in Pakistan. The report captured information about regional leaders and influencers and made suggestions for cultural touch points and possible motivators for each target audience. But the methodology was full of holes. SCL had attempted to do polling in regions using live enumerators, but the maps of rural areas they had used were incomplete, and response rates were low from people sceptical of these newcomers asking for their opinions. This produced data too incomplete or biased to be reliable. The MOD had paid some ludicrously huge sum for this, when they'd have gotten better information if they'd just hired a few locals to go into villages and ask questions.

The second problem was the way the military had chosen to disseminate its propaganda. In some projects, they had created leaflets, then dropped them all over a region. *More goddamn leaflets? The British Army was just like the Lib Dems.* And why, exactly, when there were expanding mobile phone networks throughout the country? Indeed, it was interesting how connected some of these countries were, even amid conflict. Regions without landline telephones or broadcast TV were building cell towers. I couldn't understand why western powers were ignoring this development.

More to the point, I wanted to help Nix understand that a fancy layer of psychological analysis is a complete waste if all you're going to do is dump your info out of the sky. I told him that SCL's projects would be far more effective if they focused on getting more accurate data, building algorithms, targeting specific people based on those algorithms, and using different forms of media than a leaflet or radio. Nix listened pensively, his steepled hands repeatedly tapping his mouth as he thought about what I was saying.

I also started to realise why the British and American militaries were so bad at winning the proverbial hearts and minds. This cultural and attitudinal information about the population was gathered in silos, often in contractor-led ancillary projects that were not integrated into military strategy until after primary objectives were set – in other words, the culture and experiences of local populations were an afterthought for planners, trumped by personnel and equipment. That needed to change.

AS I PONDERED WHAT DARPA and its British equivalent, the Defence Science and Technology Laboratory (DSTL), were trying to develop with their new social network and digital research programs, my mind wandered to an unexpected but not unfamiliar place for me: *fashion*. The two fields are not as disparate as it might seem. When a society jerks into extremism, so does its fashion. Think about Maoists, Nazis, Klansmen and jihadists – what do they all have in common? *A look*. Extremism starts with how people look and how society feels. Sometimes it creates literal uniforms: olive tunics and caps with red stars, red armbands, white pointed hoods, polo shirts and tiki torches, MAGA hats. These uniforms, in turn, are incorporated into the wearer's identity, transforming their thinking from *This is what I believe* into *This is who I am*. Extremist movements latch on to aesthetics because so much of extremism is about changing the aesthetics of society. Oftentimes much of what is promised is not about any tangible policies, but rather a new look and feel for a place or culture.

When I was sixteen, I dyed my hair one day to a mulberry shade of purple. There was no particular reason for choosing the colour aside from the fact it was eye-catching and I liked it. It also landed me in the principal's office, for violating the school dress code. Far from being upset or intimidated, I was totally at ease. Finally, I was talking to the principal about something other than 'disability accommodations'. I was told that I needed to change my hair back to a 'normal' colour. I refused. The principal was not happy, and the tension over my hair persisted until I left school. When I was still using a wheelchair, I spent a lot of time thinking about fit – fitting through doors, fitting in with my peers, finding clothes that fit. Computers were one passion, but fashion became another, for more than one reason. It was partly about feeling included. But it was also about being seen. As I sat at waist level, noticing the buttons, the cuts, the creases, the bulges and the folds in the clothes of classmates, I felt invisible to them. With purple hair, I was seen. And then, when the principal asked that I go back to 'normal' hair, he was ordering me to become invisible again. That was when I understood how powerful – and revealing – *a look* can be.

As I worked with the five-factor model of personality for the Lib Dems, I started to think more deeply about personality as a construct. Politics and fashion were built on the same foundation, I realised, in that they were both based on nuanced constructs of how people see themselves in relation to others. Fashion is an ideal window into personality, as choosing what to wear (or not) is a decision we all make on a daily basis. People in all cultures make choices about how to adorn their bodies, from the mundane to the extravagant. We all care about what we wear – even the straight old man from Minnesota who never wears anything but a grey T-shirt and jeans. He doesn't think he cares about how his clothes look, until you offer him a kimono or a dashiki.

I distinctly remember the final meeting I had with my university tutor at LSE, when he asked what I planned to do next. No doubt he expected to hear that I was continuing in politics or applying to a fancy corporate law firm. Instead I told him I was going to fashion school. Silence. Eyebrows raised and clearly disappointed, he unconsciously shook his head. *Fashion? As in clothes? You really want to study clothes?* But to me, fashion and politics are both, at their core, about cycles of culture and identity. To my mind, they're essentially two manifestations of the same phenomenon – a conviction that would become central to what we created at Cambridge Analytica.

Fashion has always played a role in my life, and it really was the thing that let me become more comfortable with myself. When I left school and relocated to Montréal, I was moving more and more outside of a wheelchair, but that sense of not being attractive or desirable stayed with me. Wandering around one weekend, I found myself in a vintage book shop and discovered a frayed nine-year-old issue of *Dazed & Confused* in a stack of old magazines. It was a 1998 issue, with the cover line FASHION ABLE? and showing a model with two prosthetic legs. It was guest-edited by Alexander McQueen, and inside were brilliant pictures of bodies that looked different but beautiful. After looking through that issue, I started experimenting with clothes and going out more. Montréal is the sort of place that will change you if you let it. I found myself drawn to the drag bars and admiring a form of dress that can be glam and sumptuous while mocking and upending conventional notions of beauty, bodies and gender. Drag inverted my thinking. It showed me how to not just defy these social norms, but to laugh at them and simply be who you want to be on your own terms.

In my early years in London, many of my friends were fashion students at Central Saint Martins, which is one of the constituent colleges of the University of the Arts London. I started as a student at UAL and ended up working under the supervision of Carolyn Mair, who had a background in cognitive psychology and machine learning. Dr Mair wasn't a typical fashion professor, but the match made sense, as I wasn't a typical fashion student. After I explained to her that I wanted to start researching fashion 'models' of another kind – *neural networks, computer vision and autoencoders* – she convinced the university's postgraduate research committee to allow me to commence a PhD in machine learning rather than in design. It was around this time that I also began my new job at SCL Group, so my days fluctuated between fashion models and cyberwarfare. I was keen to dive into my academic research on cultural trends, so I told Nix that I did not want to work for SCL full-time, and that if SCL wanted me, they would have to accept that I would be continuing my PhD in parallel to their projects. Nix agreed to this arrangement and SCL eventually agreed to cover my tuition fees, which felt like a godsend for me, since as an international student I had to pay the highest rate of tuition.

These two domains serve each other well, as understanding culture can equip you to unpack the dynamics of extremist movements more than purely looking at their professed ideologies can. At SCL we would watch countless numbers of radical jihadist propaganda videos, and we noticed that, beyond the violence of the clips that make it onto the news, there was a rich and well-articulated aesthetic to their style of content. Cool cars were showcased. There would be music. There was a defined masculine look to their idealised heroes, and some of the videos looked almost like clips from reality TV. The irony was that they tried to position their backward ideology as somehow modern or futuristic in a way that echoed the old Italian Futurists' promotion of a fascism for tomorrow – that it was the most expedient gateway to modernity. These films were propagating a grotesque cult of violence and hate, but beyond that, they also formed part of their culture. Their style was self-indulgent and naïvely romantic, and it bordered on kitsch. Even terrorists have pop culture.

Around this time, in September 2013, I distinctly remember thinking, *How cool is this?* I get to work in culture, but not just for someone's branding campaign. I get to work in culture for the defence of our democracy. The military just used different terms – *modelled influence attribution* or *target profiles observed acting in concert*. But in fashion, we just call that *a trend*. Dressing in concert. Hashtagging in concert. Listening in concert. Going to a concert in concert. The cultural zeitgeist itself is just *people acting in concert*. And these kinds of trends, I was sure, could be discerned in the data. Through online observation and profiling, we wanted to try to forecast these movements' adoption life cycles, their early adopters, their diffusion rates, their peaks.

In my first weeks at SCL, I began looking into how to digitise and transform the traditional tactics of information operations. This was what the firm was most interested in at the time, as it realised that there was a critical capacity gap in many NATO militaries that it could fill (and profit from) if it developed new ways of merging propaganda with ad tech. This involved exploration of what research we could draw upon about mapping out this new digital domain, such as acquiring new information sources from clickstreams and improving the targeting of narratives at their intended recipients through profiling and machine learning. There are obvious inherent complications in weaponising information. Guns and bombs kill people no matter who or where they are – the properties of physics are global. But an information weapon has to be tailored according to multiple factors: language, culture, location, history, population diversity. If you're building a non-kinetic weapon designed for scaled *perspecticide* – the active deconstruction and manipulation of popular perception – you first have to understand on a deep level what motivates people.

Insurgencies, by nature, are asymmetric, in that a few people can cause large effects. So catalysing an insurgency within the belligerent's organisation requires first concentrating resources on a few key target groups. This is optimised by good profiling and identifying the types of people who are both susceptible to new ways of thinking and connected enough to inject our counternarrative into their social network.

The most effective form of perspecticide is one that first mutates the concept of self. In this light, the manipulator attempts to 'steal' the concept of self from his target, replacing it with his own. This usually starts with attempting to smother the opponent's narratives and then dominating the informational environment around the target. Often this involves gradually breaking down what are called *psychological resilience factors* over several months. Programs are designed to create unrealistic perceptions in the targets that result in confusion and damage self-efficacy. Targets are encouraged to begin catastrophising about minor or imagined events, and counternarratives attempt to remove meaning, creating an impression of confusing or senseless events. Counternarratives also attempt to foster distrust in order to mitigate communication with others who might hamper the target's evolution. It is much harder to stay loyal to an existing hierarchy or group when you begin to think that you are being used in some unfair way, or when events seem senseless or purposeless. You become less willing to accept setbacks, take risks or comply with commands.

But simply degrading morale is often not enough. The ultimate aim is to trigger negative emotions and thought processes associated with impulsive, erratic or compulsive behaviour. This moves a target from mild or passive resistance (e.g., less productivity, taking fewer risks, rumours, etc.) into a realm of more disruptive behaviours (e.g., arguing, insubordination, mutiny, etc.). This approach has been taken in South America, for example, to provoke disunity among members of narcotics operations, increasing the likelihood of information leaks, defections or internal conflicts that erode a supply chain. The most susceptible targets are typically the ones who exhibit neurotic or narcissistic traits, as they tend to be less psychologically resilient to stressing narratives. This is because neuroticism can make a person more prone to paranoid ideation, as they tend to experience more anxiety and impulsiveness and place more reliance on intuitive rather than deliberative thinking. People high on the narcissism scale are susceptible because they are more prone to feelings of envy and entitlement, which are strong motivators of rule-breaking and hierarchy-defying behaviour. This means these targets will be more likely to develop an exaggerated suspicion of harassment, persecution, victimhood or unfair treatment. This is the 'low-hanging fruit' for initiating the subversion of a larger organisation. Later, this learning would serve as one of the foundations for Cambridge Analytica's work catalysing an alt-right insurgency in America.

Let's be clear: these operations are not some kind of therapeutic counselling; they are a form of psychological attack. It's important to remember that, in a military context, the target's personal agency or consent is not a concern. The target is the enemy. The choice for the military is often either to send in a drone to incinerate the enemy or to mess with the enemy's unit to such an extent that they begin to fight among themselves or get sloppy and make exploitable mistakes. If you are a military commander or an intelligence officer, psychological manipulation is the 'light touch' approach.

With the advent of social media, suddenly military and security agencies had direct access to the minds and lives of guards, clerks, girlfriends and runners of criminal and terrorist organisations all around the world. What the social data offered was a trail of detailed personal information that previously would have taken months of careful observation to gather. The targets were in effect creating their own dossiers with rich data that could quicken a psychologist's assessment of their disposition. This spurred a host of research into psychological profiling that could be automated with machine-learning algorithms. These algorithms would allow agencies to widen their net through automation and reach the scale of an old-school leaflet drop, but with the precision of targeted messages. In 2011, DARPA began funding research into psychological profiling of social media users, how anti-government messages spread, and even online deception. Engineers at Facebook, Yahoo and IBM all participated in DARPA-funded research projects to assess how information is consumed and spread. The Russian and Chinese governments also launched their own social media research programs.

On my very first day at SCL, Nix asked whether I'd heard of a company called Palantir. He'd learned about it from an unusually well-connected SCL intern named Sophie Schmidt – the daughter of Eric Schmidt, a billionaire and then executive chairman of Google. A few months earlier, as she was finishing up her internship, she'd introduced Alexander to some of the executives at Palantir. Co-founded by Peter Thiel, a well-known venture capitalist in Silicon Valley who was also an independent director of Facebook, Palantir was a massive venture-capital-funded company that undertook information operations for the CIA, the National Security Agency, whose mission is to analyse signals intelligence and data for national security purposes, and the Government Communications Headquarters (GCHQ), the British counterpart to the NSA. Nix was obsessed. He wanted SCL to do what Palantir was doing.

In my first few months, I worked on small pilot projects in various countries with Brent Clickard, the Cambridge-based psychologist, and a friend of his named Tadas Jucikas. I first met Jucikas at the Royal Automobile Club, a private members' establishment with bars, squash courts and billiard rooms, where the upper crust go to socialise and do business. It was founded back in 1897, when driving a car was an exorbitantly expensive hobby, and has retained its air of lavish elegance. When I walked up to the colonnaded clubhouse, I saw Jucikas standing in the lobby beside a bright-red antique racing car, his eyes hidden behind tortoiseshell sunglasses. He was wearing a beautifully cut herringbone jacket with a crisp pocket square. It was all so extra, but I was into it.

He led me inside the club, where we drank a couple of boulevardiers before moving to the balcony to smoke cigars. Jucikas had grown up in rural Lithuania, where he watched Soviet tanks roll through his town as a boy. He was obviously brilliant. As we sat on the balcony, savouring our cigars and discussing artificial intelligence and data pipelines, Jucikas opened a satchel and pulled out a diagram he'd created. Clickard had told him about the scope of some of the projects, so Jucikas had mapped out a proposed data science pipeline on how to ingest, cleanse, process and deploy data from online profiles of people. He had been doing his PhD research on modelling and predicting the behaviour of *C. elegans* roundworms and said that he simply swapped out the worms for people. Jucikas proposed pulling a wide variety of data by building automated data-harvesting utilities, using algorithmic imputations to consolidate different data sources into a single unified identity for each individual, and then using deep-learning neural networks to predict our desired behaviours. We would still need a team of psychologists, he said, to create the narratives needed to change behaviours, but his pipeline served as the first sketch of the targeting system. But what I loved most was that he colour-coded it to make the journey look like the London tube map. As he went through his explanation, it was clear that he was perfect for the job.

So it was that Clickard, Jucikas and I began working together, and I eventually persuaded Mark Gettleson to join as well. Suddenly I was surrounded by a team of impeccably dressed, blazingly smart, impossibly quirky individuals. And Nix was the ringleader, the grinning, soulless salesman who didn't understand anything we were doing but wasted no time selling it fast and hard to anyone he thought might pay. He ruled the office with haughty proclamations and crude sexual jokes.

There was an 'anything goes' atmosphere at SCL, perfectly crystallised in a moment that took place a couple of months after we all started. Normally I dress in T-shirts and hoodies, but one afternoon I came into the office after a London Fashion Week event wearing a vibrant burgundy Prada jacket with matching high-waisted trousers and cream-coloured Dr Martens shoes printed with skulls and roses in the style of tattoo flashes. Nix took one look and said, 'Chris, what the fuck are you wearing?'

To which Brent answered, 'We fight terror in Prada.'

GETTING THE REQUIRED DATA to build Jucikas's envisioned targeting system would not be easy, but it was possible, due to a fluke of history in some parts of the developing world. Although there was substantial underdevelopment of traditional telecommunications infrastructure, largely stemming from corruption and the neglectful legacies of colonial administrations, some of the world's poorest countries had leapfrogged generations of technology, achieving impressive advances in mobile networks.

In Kenya, for example, local laws and customs made it difficult for some people to get a bank account, leading to a system in which Kenyans used cash to buy mobile phone credits, which could then be traded as a kind of digital currency. In fact, we found that people in many poorer nations distrusted banks, having lived through economic crises, hyperinflation and bank collapses, and used the same mobile workaround. This setup meant that everybody needed a phone, and that it needed to work well, so that in otherwise impoverished nations, there'd been rapid investment in relatively decent mobile infrastructure.

One unintended consequence of having large pluralities of citizens connected via mobile phone networks was that everybody could be traced, tracked, profiled and communicated with. Jihadist networks such as ISIS, AQAP and Boko Haram had already figured this out, taking advantage of easy access to the minds of future conquests. And that turned the rules of warfare upside down.

Next we needed a case study – a location where we could scale to a nation-state level, to show potential military clients what we were capable of doing. Trinidad and Tobago, with 1.3 million people, fit the bill perfectly. It was an island nation, self-contained yet with a variety of cultures. There was an Afro-Caribbean population, an Indo-Caribbean population and a smattering of white people, creating an interesting cultural tension to explore. It was an ideal laboratory in which to run our experiments at scale.

The Trinidad Ministry of National Security wanted to know whether it was possible to use data to identify Trinidadians who were more likely to commit crimes – and, beyond that, whether it was possible to predict when and how they might do it. SCL had a long history of operating throughout the various micronations of the Caribbean, and after it helped selected politicians get into power, the firm would often recoup its investment in government contracts. At SCL, we started referring to this as the Minority Report Project, after the Philip K. Dick story (adapted for film by Steven Spielberg) in which a futuristic PreCrime Division arrests people before they're able to commit crimes. But the truth was, the Trinidadian government wasn't interested only in reducing crime. They knew that if we built a tool to forecast behaviour, they could use it in elections. They weren't just focused on future criminals; they also wanted to zero in on future political supporters.

The team anticipated vast swathes of data, because senior Trinidad government contacts were offering SCL access to the unredacted, de-anonymised census – in the developing world, privacy is a concern usually reserved for the rich. Essentially the Trinidadian government was violating the privacy of all its citizens in one swoop.

The raw census data would obviously be useful for the project, but it wasn't a resource we could expect to have available to us in developed countries. SCL needed to explore using the internet to collect relevant data, to create a tool that would be cross-culturally and cross-nationally applicable. So SCL's next step was to send people on the ground to Caribbean telecom companies, to ask if SCL could tap into their data 'firehose' in real time. To my surprise, this was possible.

Working with a set of contractors, SCL was able to tap into the telecom firehose, pick an IP address, and then sit and watch what a person in Trinidad was browsing on the internet at that very moment. Not surprisingly, it was a lot of porn. People were browsing everything imaginable, including the culturally specific *'Trini Porn'*. I can remember sitting around the computer one evening and watching as someone toggled between looking up plantain recipes and watching porn, all while Nix laughed at them. It was a revoltingly giddy laugh, almost infantile. He looked up the IP address and then opened up Google Maps satellite view to see the neighbourhood this person lived in.

As Nix watched the screen, I began to watch him, taking such deep, nasty pleasure in the chance to ridicule and exploit others. It was classic

Nix – or 'Bertie,' as his pompous peers called him. Like many Old Etonians, he excelled at banter, flirtation and entertainment. The directors of SCL assigned him to lead the firm's side business of rigging elections in forgotten countries of Africa, the Caribbean and South Asia. It was with cabinet ministers of micronations that Nix was completely in his element. Performing the role of the English gentleman, he would give these politicians access to anything they wanted in the old imperial capital of London – the prestigious clubs frequented by royals and prime ministers, invitations to exclusive parties, or, if desired, the private company of elegant and open-minded women.

Nix preyed upon the colonial fetishes and insecurities of the men who ran the nations of the empire. Once he gained their trust, he would then broker deals between ministers who were looking for validation and women, and businessmen who were looking to exploit corrupt business opportunities and travel unnoticed. Sovereignty, Nix learned, was an extremely valuable commodity. Even the smallest and most obscure island nation could offer two things of exceptional value: passports and tax immunity. He had inherited tens of millions of pounds and never needed to work. He could have dedicated his life to noble pursuits or simply settled into a life of leisure, sponging off his trust fund. But instead he chose SCL. Nix couldn't help himself – he was intoxicated by power. Born too late to play colonial master in the old British Empire, he treated SCL as the modern equivalent. As Nix put it in one of our meetings, he got to 'play the white man'. 'They [are] just niggers,' he once said to a colleague in an email, referring to black politicians in Barbados.

We were spying, pure and simple, with cover from Trinidadian leaders. It felt bizarre – unreal – to be observing what people were watching on a tiny, faraway island, somehow more like we were playing a video game than intruding on the private lives of actual people. Even today, thinking back on it, Trinidad seems more like a dream than something we actually did.

But we did do it. The Trinidad project was the first time I got sucked into a situation that was grossly unethical, and, frankly, it triggered in me a state of denial. As I watched those livestreams, I didn't allow myself to actually picture the human prey, people who had no idea that their private behaviour was delighting sinister audiences half a world away. The Trinidad project was my first taste of this new wave of digital colonialism. We arrived unannounced with our superior technology and moral disregard, no better than the king's armies. Except this time, unlike the conquerors of old, we were completely invisible.

ALREADY, IN MY EARLY months of working with Nix, it had become clear that he had no real business ethics – or personal ethics, for that matter. He seemed willing to go to any lengths to win a project, and he'd peacock around the office bragging about this deal or that deal. He described everything in terms of sexual conquest: in the early stages of negotiations, the two sides were 'feeling each other up' or 'slipping in a finger'. When a deal closed, he'd exclaim, '*Now* we're fucking!'

In August 2013, not long after the Rabaa massacre, representatives of the Egyptian government came to London for meetings. This was one of the early movements where social media and instant messaging on mobile apps played a significant role in mobilisation. The Egyptians we met were interested in using our information programs to combat what they called 'political extremists'. Several scenarios were discussed about how to create havoc within the movement, including starting rumours to spread through mobile messaging or riling up crowds with planted confederates and arresting protestors. This wasn't the kind of project I'd expected SCL to take on, and I felt morally opposed to what they were asking us to do. It was here that I was confronted with the very subjective meaning of counter-extremism. It seemed entirely hypocritical to, on the one hand, frustrate jihadist groups in places like Pakistan and then, on the other, assist an autocratic and Islamist-backed regime in Egypt in creating its own tyranny of people. But Nix didn't care. Business was business; he just wanted to clinch the deal.

The main challenge for me and the growing team of psychologists and data scientists at SCL was in the objective substance of extremism itself. What does it mean to be an extremist? What exactly is extremism, and how can you model it? These were subjective definitions, and clearly the Egyptian government had one idea, while we had another.

But if you want to be able to quantify and predict a trait, you have to be able to create a definition of it. We went around and around, discussing the question in theoretical terms, but the reality of it felt sobering: *Extremism is whatever you want it to be.* In the end, SCL didn't undertake the project, so I just compartmentalised my concerns and kept working.

I began trying to avoid Nix at the office – everyone did, because he behaved so repulsively. His efforts to take me under his wing – to remake me in his image – were a dismal failure. Our backgrounds were too different, for starters. Even if I didn't find Nix's arrogance and snobbery appalling, I never could have disguised myself as a 'respectable' Old Etonian, and his constant hectoring – what to wear, how to speak, etc. – only made me more self-conscious. We did occasionally bond over a mutual fondness for good whisky, but mostly I kept my distance.

The projects that most engaged me were those that were doing some good in the world, such as programs to de-radicalise what the military affectionately called the YUMs – *young unmarried males* – in the Middle East and root out jihadist behaviour. I found ways to justify staying, reasoning that even if Nix was obviously a villain, there were still lots of good people working for SCL. I decided just to keep my head down and keep working.

In late 2013, I was asked to join a meeting with a potential client in an African nation. This would be a political project, I was told, involving targeting voters ahead of an upcoming election. I didn't know much about the country, but I assumed we could get hold of the necessary data through mobile networks or public sources, so I said 'Sure'. We met with the client, who turned out to be the country's minister of health, at an expensive restaurant in London.

At first, the discussion ran more or less as I'd expected. We talked about what services the client needed and how SCL could provide them. Then the conversation turned to how the project could be funded, and the firm made a proposal: the client could leverage an existing multi-million-dollar project for the country's ministry of health, and SCL would quietly be added as a subcontractor, dipping into the project's budget to conduct political research. Afterward, another staff member followed up with an email: 'The health component of a larger survey will act as a prelude to an election campaign,' and noted, 'the political component has also been approved.' The email went on to explain that the ministry of health survey would include questions about voting behaviour and support for the current administration. Of course, using taxpayer funds from the ministry of health for political campaigns is unlawful.

I said nothing during the meeting, but afterward I went to see Alexander. 'This can't be legal,' I told him. To which he replied, 'You can't expect anything legal with these people. It's Africa.'

Nix was extremely good at getting people to doubt themselves, and throughout my time at SCL, I kept falling for it. Other times, he was less convincing. One time he took me onto the roof, high above our new offices on New Bond Street, for a 'man-to-man,' where he offered me a horse if I helped win a project. He had lots of horses, apparently. I said I didn't want one. 'Oh, right,' he said. 'A pony for you, then.' After speaking to him, I often wasn't sure what I was supposed to be feeling – offended by what he'd said or embarrassed at my naïveté.

I couldn't believe that the African project would play out as planned, but it did. SCL created a proposal for a subcontract and submitted it for approval with the ministry of health. Over a period of many months, as the health-related projects proceeded, some of the money – *millions of dollars* – did not actually go to ministry of health programs. It was split between the minister's political campaign and SCL, with SCL's cut arriving from the country's embassy in diplomatic bags so that they could bypass any border inspection or declaration. I removed myself from the project early on, recognising that it was morally and legally beyond the pale.

The deeper I got into SCL's projects, the more the office culture seemed to be clouding my judgment. Over time, I was acclimatising to their corruption and moral disregard. Everyone was excited about the discoveries we were making, but how far were we willing to go in the name of this new field of research? Was there a point at which someone would finally say *Enough is enough*? I didn't know, and in truth, I didn't want to think about it. And that was when Nix sent me an email: 'I'd like you to meet Steve from America.'

4

STEVE FROM AMERICA

I MUST HAVE DOZED OFF, BECAUSE THE CONDUCTOR'S ANNOUNCEMENT startled me: *'Alight here for Cambridge!'* It was October 2013 and I'd woken up that morning at 5:00 to make the 6:40 train out of London's King's Cross Station. Nix had booked me on the early one to save himself five pounds. I jumped out of my seat and accidentally knocked into the elderly lady beside me. She just glared, clutching her purse, as English people do. I was running out, looking back to say sorry, when I tripped. *'Mind the gap!'* Too late.

I stood up, only to realise that I'd somehow misplaced my wallet, and then watched in dread as the train slowly pulled out of the station. *Ugh*. Without cash or cards, I called Nix and asked him to book a prepaid taxi. 'Walk there,' he said. 'You should have been more careful.' I was too tired to argue, and he was clearly in a mood, so I did as Nix said – I walked, departing the station into the mist and drizzle of an early October morning. Cambridge was just starting to wake up.

With several hours to kill before my appointment, I meandered through Parker's Piece, a small common green, as student-athletes began a morning practice against a backdrop of a church spire peeking through the trees. From there, I walked through the town's winding medieval stone streets, past small shops and the towering walls of England's second-oldest university, dating back to the year 1209. After continuing to Thompson's Lane, near the River Cam, I arrived at the small but clearly expensive Varsity Hotel.

Working with a military contractor, I met all sorts of bizarre characters, most of whom had a strong desire for 'absolute discretion' – it wasn't the least bit unusual to not know the full identity of a person before the first meeting. The day before, Nix had come into the office looking slightly agitated and walked immediately over to me, where he put both his hands on my desk and leaned into my face. 'I need you to meet someone tomorrow in Cambridge,' he said. 'I can't get into his head, but I think you can.'

I asked who I'd be meeting.

'I'll email you the details later.'

In this case, the extremely unhelpful instruction I got from Nix was simply to meet *'Steve from America,'* with no details beyond a request that I 'bring data'.

I sat alone in the hotel lobby for an hour before texting Nix, asking for Steve's number. He read the text but didn't respond. After another fifteen minutes, this gruff character walked up and looked me over.

'You the guy?' he asked.

'Yeah, I am,' I replied. Based on the clients SCL normally had, I'd expected some government or agency type. Instead I found myself looking at a dishevelled man wearing two collared shirts, as if he'd forgotten to take one off before putting the next one on. He was unshaven, with greasy hair and that layer of grime you get from a transatlantic flight. His eyes showed flecks of bright red that matched the web of rosacea on his skin. In all, the vibe he gave off fell somewhere between used-car salesman and madman. He looked tired or dazed; I assumed it was just jet lag.

The lift was a classic English setup that barely had room for two, meaning I had to work hard not to touch this guy. I was wearing monochromatic Dries Van Noten – dark navy suit trousers with a matching overshirt that blended like an obliquely cut Maoist jumpsuit.

'You weren't what I'd imagined,' he half joked. *Yeah, you aren't such a looker yourself, hun . . .*

He was staying in a suite on the top floor. Save for the bold wallpaper on an accent wall, the décor was minimal and modernist, which made for a stark contrast with the panoramic view of the medieval city below. The absence of luggage seemed weird, but not worth dwelling on. Then I hesitated. *Oh, wait, I'm alone in a posh hotel room with some old guy.* I looked over at the king bed, then noticed a little bottle of hand lotion on the table next to it. *Fuck, fuck, fuck* – was Nix using me as bait?

I clutched my bag, hoping the laptop inside was heavy enough to land an effective blow. At that moment, Steve Bannon walked over to the large sofa adjacent to the bed and offered me a seat. To my extreme relief, he grabbed a chair for himself and asked if I wanted some water. As he sat down, his stomach spilled over his waistline.

'Nix tells me you're doing research on cultural change,' he said. 'Tell me about that.'

I told him we were using computers to quantify cultural trends and predict how they will evolve in places at risk for extremism. 'We try to glimpse into the destiny of cultures,' I said, aiming to distill decades' worth of computational and social theory. Bannon rolled his eyes. 'Yeah, yeah, yeah. You can cut the bullshit and tell me what it is you *actually* do.'

We talked for four hours – not only about politics but about fashion and culture, Foucault, the third-wave feminist Judith Butler, and the nature of the fractured self. On the surface, Bannon seemed utterly predictable – another old, straight white guy – but he spoke with a certain wokeness I hadn't expected at all. In fact, I quickly decided he was kind of cool. As we started trading ideas on measuring culture, I offered to show him some of our data. I opened a Tableau workbook and called up a map of Trinidad. I clicked a button and a layer of neon-yellow dots began to populate the map. 'Those are real people, by the way,' I said. 'They are the ones we have demographic data on … gender, age, ethnicity.'

I clicked again and more dots appeared. 'And now we are adding in online footprinting – like internet browsing.'

I clicked again. 'And here are records with census information … and now social media profiles.' I continued to add layers and he leaned in. The map lit up more and more, with little clusters of dots growing outward until, after the final click, the map was dazzling, in a multitude of colours. He asked who had paid for it, but I told him I couldn't say. As I started to outline the types of research into social media networks that DARPA was funding, he asked if something similar could be carried out in America.

'I don't see why not,' I said.

STEVE BANNON WAS BORN in Virginia in the early 1950s to a working-class Irish Catholic family. He went to a Catholic military high school and graduated with a degree in urban affairs from Virginia Tech, then served in the Navy as a surface warfare officer before a post at the Pentagon writing reports on the status of the US Navy fleet worldwide. In the 1980s, his life took an academic turn – a 1983 master's in national security studies from Georgetown University, a 1985 MBA from Harvard Business School. After a tour in investment banking, Bannon moved on to making films in Hollywood as an executive producer, director and writer. He worked on more than thirty films, including a documentary about Ronald Reagan. In 2005, Bannon joined the Hong Kong-based Internet Gaming Entertainment (IGE), and a year later he brought in a $60 million investment, half of which came from his former employer Goldman Sachs. The company eventually rebranded as Affinity Media Holdings, and Bannon continued to help run it until 2012, when he joined Breitbart. Next, Bannon co-founded the Government Accountability Institute, which eventually published the book *Clinton Cash,* by Breitbart News editor-at-large Peter Schweizer.

In 2005, the right-wing commentator Andrew Breitbart began Breitbart.com, an online news aggregator, and by 2007 it had grown to publish original content as Breitbart News. The site ran on the undercurrent of Breitbart's personal philosophy, which has been referred to as the Breitbart Doctrine: politics flows from culture, and if conservatives wanted to successfully dam up progressive ideas in America, they would have to first challenge the culture. And so Breitbart was founded to be not only a media platform but also a tool for reversing the flow of American culture.

When Andrew Breitbart (who had introduced the Mercers to Bannon) died suddenly in 2012, Bannon took his place as senior editor, and assumed his philosophy. At our first meeting, he was the executive chair of Breitbart and had come to Cambridge in search of promising young conservatives and candidates to staff his new London bureau. The logic, as we later learned with Brexit, was that Britain served as an important cultural signifier for Americans. Win the Brits, and so falls America, Bannon later told me, as the mythologies and tropes of Hollywood had crafted an image of Britain as a country of educated, rational and classy people. He had a problem, though. For all the site's sound and fury, it became pigeonholed as a place for young, straight white guys who couldn't get laid. Gamergate was one of the first, most public instances of their culture war: when several women tried to bring to light the gross misogyny within the gaming industry, they were hounded, doxed and sent numerous death threats in a massive campaign against the 'progressives' imposing their 'feminist ideology' onto gaming culture.

Gamergate was not instigated by Breitbart, but it was a sign to Bannon, who saw that angry, lonely white men could become incredibly mobilised when they felt that their way of life was threatened. Bannon realised the power of cultivating the misogyny of horny virgins. Their nihilistic anger and talks of 'beta uprisings' simmered in the recesses of the internet. But growing an army of 'incels' (involuntary celibates) would not be sufficient for the movement he fantasised about. He needed to find a new approach.

This is one of the odder moments in the Cambridge Analytica saga – the random airplane conversation that changed history. Several months before I met Bannon, two Republican consultants, Mark Block and Linda Hansen, happened to be sitting next to an ex-military officer who had worked as a subcontractor for a company that utilised 'cyberwarfare' in elections. Block fell asleep on the flight, but Hansen and her seatmate started chatting, and the man told her about SCL's projects in information warfare. When the flight landed, Hansen told Block they needed to contact Nix. Block, who had been the campaign manager for Herman Cain, was well connected to the fringe elements of Republican circles. He knew Bannon and understood immediately that SCL would be of interest to him. So Block connected Bannon with Nix, and I wound up in this hotel suite meeting the man who would later stage a mass manipulation of the American psyche.

By the time I walked through the doors of the Varsity Hotel, Nix had already met with Bannon several times in New York. But when Nix tried to explain our projects, he ran into a problem – he didn't actually understand what we were working on. He was in deeply unfamiliar territory with Bannon, who cared more about the details of the research than the pedigree of the researchers. Inside SCL, Nix was typically relegated by the other directors to deal with their 'less serious' clients. Nix became more active in the company after his father, who was a large shareholder, died in 2007. He had graduated with average marks in art history from the University of Manchester but preferred the various enterprises of wealthy friends and family to galleries or libraries.

Bannon was not a typical client for Nix, who was far more used to dealing with ministers or businessmen from the developing nations of Britain's old empire. Bannon did not need a second passport from a tropical nation. He was not looking for colonial cosplay in London, and he did not care how Nix pronounced his words or about the tailoring of his bespoke suit. Bannon wanted *real* things. It was deeply disorienting for a man accustomed to seducing ministers with scantily clad Ukrainian women and inebriated Etonian banter.

Originally, Nix suggested to Bannon that we meet somewhere on London's Pall Mall, a street lined with grand stone buildings. A couple of blocks north of Buckingham Palace, Pall Mall begins at Trafalgar Square and ends at St James' Palace, the sixteenth-century residence of several members of the royal family. The area is home to some of Britain's most exclusive private gentlemen's clubs, where black tie is common and Nix socialised with his peers, sucking down drinks in opulent surroundings. Nix had imagined an elaborate dinner in a private dining room at the Carlton Club, meticulously planning the menu and serving staff, only to be rebuffed at the last minute.

Still, Nix knew that everyone, including Bannon, suffers the yearning of an unfulfilled secret self. He realised that the American was lounging in the ancient universities of England to play out a role – when Bannon looked in the mirror, he saw a philosopher. To win him over, Nix would need to help him achieve his fantasy of becoming a thinker of big thoughts. And so my 'academic' vibe became just what he needed to lure Bannon into role-play.

Today Bannon is famous, but as we sat in that hotel room in the autumn of 2013, I knew virtually nothing about *Steve from America*. Even so, I quickly realised we were kindred spirits. We had ended up in politics, but our shared passion was culture, with his ambitions in film and mine in fashion. He indulged my interest in deconstructing trends and agreed that many of our social norms could be boiled down to aesthetics. And we both saw what was bubbling in tech and online spaces. He talked about gamers, memes and MMORPGs – online games like *World of Warcraft* with huge numbers of players. He used the word 'pwned' in a sentence, which is a gamer expression that implies domination or humiliation of a rival. We connected on all the things that made us weird. As we sat talking together, I found myself growing unexpectedly comfortable with him. He was no political hack, but a fellow nerd given permission to speak freely.

When Bannon said he was interested in changing culture, I asked him how he defined culture. There was a long pause. I told him that if you can't define something, you can't measure it, and if you can't measure it, you can't know if you are changing it.

Rather than dive deep into theory, I gave Bannon a grossly simplified example of what culture is by using cultural stereotypes. Italians have a reputation for being more passionate and extroverted than other people. (Having dated one, I can testify to the grain of truth behind this reputation.) And while it's obvious that not all Italians are loud and brimming with passion, if you visit Italy, you'll probably find more people who are extroverted in their presentation than if you visit, say, Germany or Singapore. This can be thought of as a *norm* – the peak on a bell-shaped distribution curve of extroversion or loudness. And perhaps Italy peaks a bit further up the scale than other countries.

When we describe cultures, we use the language and vocabulary of personality. We use the same words to describe both *people* and *peoples*. On the one hand, we can't stereotype at the individual level, because every person is different. But on the other hand, we can say that, in a broader sense, Italian culture can be characterised as probably a bit more outgoing than many other cultures.

If we can measure or infer certain traits in individuals using personal data, and then use those same traits to describe a culture, we can chart a distribution, creating an approximate metric for that culture. This framework made it possible for us to propose how we could use personal data found on social media, in clickstreams, or from data vendors to identify, for example, who the most extroverted Italian people are through their patterns of behaviour as individual consumers and users. Then, if one wants to shift the culture to make it slightly less extroverted, this data gives us a list of actual named Italians, ordered by their degree of extroversion, whom we could track and target over time, trying to chip away at their extroversion. In other words, culture change can be thought of as nudging the distribution curve of culture up or down. What the data allowed us to do was to disaggregate that culture into individuals, who became movable units of that society.

Bannon was someone who liked to talk, but when I got into a subject that interested him, he was quiet and even deferential. But he was also eager to get back to applications. To understand how this might become a practical campaign, think of public health. When a communicable disease threatens a population, you immunise certain vectors first – usually babies and old people, as they are most susceptible to infection. Then nurses and doctors, teachers and bus drivers, as they are most likely to spread a contagion through wide social interaction, even if they do not succumb to the disease themselves. The same type of strategy could help you change culture. To make a population more resilient to extremism, for example, you would first identify which people are susceptible to weaponised messaging, determine the traits that make them vulnerable to the contagion narrative, and then target them with an inoculating counter-narrative in an effort to change their behaviour. In theory, of course, the same strategy could be used in reverse – to foster extremism – but that was not something I had even considered.

THE GOAL IN HACKING is to find a weak point in a system and then exploit that vulnerability. In psychological warfare, the weak points are flaws in how people think. If you're trying to hack a person's mind, you need to identify *cognitive biases* and then exploit them. If you walk up to a random person on the street and ask, 'Are you happy?'

the chances are high that she will say yes. If, however, you walk up to that same person and first ask, 'Have you gained weight in the last few years?' or 'Are any people from your high school more successful than you?' and *then* you ask 'Are you happy?' – that same person will be less inclined to answer yes. Nothing about her personal situation or history has actually changed. But her *perception* of her life has. Why? Because one piece of information in her mind was weighted more than the others.

What we played with as the questioner was how she was weighting that information, which in turn affected her judgment of that information. We biased her mental model of her life. So which is true? Is she happy or not happy? The answer depends on which information is being pulled to the front of her mind. In psychology, this is called priming. And this is, in essence, how you weaponise data: you figure out which bits of salient information to pull to the fore to affect how a person feels, what she believes, and how she behaves.

Unless someone's parents are secretly Vulcan, no one on earth is a purely rational thinker. We are all affected with cognitive biases, which are the commonly occurring errors in our thinking that generate flawed subjective interpretations of information. It is completely normal for people to process information with bias – in fact, everyone does – and oftentimes these biases are harmless in day-to-day life. These biases are not random in each person. Rather, they are systematic errors, meaning they create patterns in common forms of irrational thinking. In fact, thousands of cognitive biases have been identified in the field of psychology. Some biases are so common and seemingly intuitive that it can be hard for people to even recognise that they are actually irrational.

For example, the psychologists Amos Tversky and Daniel Kahneman conducted a study that asked participants a very simple question: '*Suppose you sample a word at random from an English text. Is it more likely that the word starts with a* k, *or that* k *is the third letter?*' Most people responded with the former, that words that start with *k* (e.g., *kitchen, kite* or *kilometre*) are more likely. However, the opposite is true, and one is actually twice as likely in a typical English text to encounter words where the third letter is a *k*, such as *ask, like, make, joke* or *take*. They tested for five letters (*k, l, n, r* and *v*) like this. It is easier for people to think of words by first letter because we are taught to organise (or alphabetise) words by their first letter. However, people conflate this ease of recall with frequency or probability, even when this is far from the truth. This cognitive bias is called the *availability heuristic,* and is just one of many biases that affect our thinking. The bias is why, for example, people who see more news reports of violent murders on the news tend to think that society is becoming more violent when in fact global murder rates have been declining overall during the last quarter century.

I had been pondering these ideas based on my experiences in politics, then fashion and then information warfare. Political extremism, for example, is a cultural activity with parallels in fashion: they're both based on how cultural information proliferates through the nodes of a network. The rise of jihadism and the popularity of Crocs can both be thought of as the products of information flows. When I started my research into cultural information for SCL's counter-extremism work, I drew upon similar concepts, approaches and tools to those I was exploring in fashion forecasting – adoption cycles, diffusion rates, network homophily, etc. The work was all about trying to anticipate how people would internalise and then spread cultural information – whether that meant in joining a death cult or in choosing a wardrobe.

Bannon immediately grasped all of this, even telling me that he believed, as I do, that politics and fashion are essentially products of the same phenomenon. It was obvious that he treated intelligence gathering in a broad and deep way, which is not something I've seen many people in politics do. And that's what makes him so powerful. He reads about intersectional feminism or the fluidity of identity not, as I later learned, because he's open to those ideas but because he wants to invert them – to identify what people attach themselves to and then to weaponise it. What I didn't know that day was that Bannon wanted to fight a cultural war, and so he had come to the people who specialised in informational weapons to help him build his arsenal.

Bannon and I were clearly on the same wavelength, and the conversation that day flowed so naturally, it felt as if we were flirting – but not, because that would be gross. But intellectually, we were a match. I left that meeting feeling uplifted and validated by someone who had taken the time to listen. Bannon came across as a reasonable guy when I first met him – nice, even. I could tell he appreciated learning new ideas and got excited by their possibilities. But what struck me was how this guy was a cultural maven and a tech nerd. I realised he had a bit of a libertarian streak, but we hadn't talked that much about politics.

Then I remembered I had lost my wallet. I called Nix to tell him how everything had gone – and that I needed a new ticket. 'Chris, I'm busy, sort it out yourself.'

BANNON'S INTEREST IN OUR work wasn't merely academic; he had big ideas for SCL. He told Nix of a major right-wing donor who might be persuaded to make an investment in the firm. Robert Mercer was unusual for a billionaire. He'd gotten a PhD in computer science in the early 1970s, then went on to become a cog in the wheel at IBM for twenty-some years. In 1993, he joined a hedge fund called Renaissance Technologies, where he used data science and algorithms to inform his investments – and made a stupid amount of money doing it. Mercer wasn't one of these wheeler-dealer types who frenetically bought and sold businesses. He was an extremely introverted engineer who applied his technical skills very specifically to the art and science of making money.

Over the years, Mercer had donated millions of dollars to conservative campaigns. He also started the Mercer Family Foundation, run by his then-thirty-nine-year-old daughter, Rebekah, which originally supported research and other charities, but had begun to also donate to politically involved nonprofit groups. His wealth and influence placed him alongside the Koch brothers and Sheldon Adelson in the pantheon of Republican donors. The news that Mercer might be willing to invest in SCL made Nix salivate. Mercer's profile was one of disrupting the financial sector. Renaissance was one of the highest-performing hedge funds in the industry – and Mercer built the firm by eschewing traditional finance backgrounds and instead hiring physicists, mathematicians and scientists to build his firm's algorithms. But Mercer, it seemed, wanted us to attempt an even more ambitious version of profitable disruption. By profiling every citizen in a country, imputing their personalities and unique behaviours, and placing those profiles in an *in silico* simulation of that society (one created inside a computer), we would be building the first prototype of the artificial society. If we could play with an economy or culture in a simulation of artificial agents with the same traits as the *actual people* they represented, we could just possibly create the most powerful market intelligence tool yet imagined. And by adding quantified cultural signals, we were verging on a new area of something akin to 'cultural finance'. We thought that if we got it right, we could run simulations of different futures of whole societies. Forget shorting companies; think about entire economies.

It turned out that what Mercer had in mind went beyond the economy, but at the time our focus was on demonstrating what SCL was capable of doing. After some deliberation, Bannon decided that we should run a proof of concept in Virginia, which felt like a good microcosm of America. It's a little bit northern and a little bit southern. It has mountains and coastal areas, military towns, wealthy DC suburbs, rural areas and farms, and a cross section of rich and poor, black and white. The Virginia experiment would mark the first time we'd played with data in the United States. As I'd done with the LPC and the Lib Dems, we started off with qualitative research – unstructured, open-ended conversations with local people. Nobody on the SCL team was American, and we didn't know anything about Virginia, which was as foreign to me as Ghana. The obvious first step was to visit the state and talk to people, to learn how they perceived the world and what mattered to them. We couldn't generate questions until they had introduced themselves to us, in their own way and in their own environment. Once we had a better feel for what Virginians cared about and how they approached things, we could then structure specific questions for quantitative research. Politics and culture are so intertwined that one cannot usually study one without the other.

So, along with Mark Gettleson, Brent Clickard and a few others, I flew to America, arriving in Virginia in October 2013, shortly before statewide elections there. One of the things we heard in focus groups was concern about the Republican candidate for governor, the former state attorney general Ken Cuccinelli. He was a super-right-wing type who had advocated for initiatives to roll back gay rights and fight environmental protections. The Republican Party in Virginia has an enormous bloc of evangelical Christian voters, and Cuccinelli needed them if he was going to win. But, as we discovered in our research, he went so hard after their votes, he overshot his mark.

One of Cuccinelli's initiatives was to petition a federal court to reverse its ruling on Virginia's 'Crimes Against Nature' law. Originally passed in 1950 and formally struck down in 2013 by the US Fourth Circuit Court of Appeals (in light of a 2003 US Supreme Court decision to decriminalise sexual activity between consenting adults), the statute technically outlawed oral and anal sex. Cuccinelli argued that the law was needed to combat pedophilia. On paper, he reminded me of crazed politicians we'd encountered in parts of Africa, obsessed with gays and their bedroom sins. But social extremists and weirdos can be found anywhere, even in white-bread America.

People in our focus groups – particularly straight, red-blooded American men – kept saying how weird they thought this was. Ban the gay stuff, sure, but why ban *all* non-procreative sex? Why was Cuccinelli so opposed to blow jobs? Let's be real – isn't that a little weird? These guys kept talking about how they didn't like thinking about Cuccinelli and getting head, and who could blame them? We kept hearing about this issue, so we decided to try an experiment.

In the five-factor model of personality, conservatives tend to display a combination of two traits: lower openness and higher conscientiousness. In the most general way, Republicans aren't likely to seek out novelty or to express curiosity about new experiences (with closet cases being the obvious exception). At the same time, they favour structure and order, and they don't like surprises. Democrats are more open but also often less conscientious. This is in part why political debates often centre around behaviour and the locus of personal responsibility.

Our qualitative research told us, among many other things, that Virginia Republicans were put off by Cuccinelli's obsession with blow jobs. And psychometric testing also told us that Republicans don't like unpredictability. Could we create a strategy, using those two observations, to move the needle of opinion on Cuccinelli?

This was where Gettleson's brilliance came into play. He was particularly fascinated by alpha-male voters and Cuccinelli's conundrum with them, but he also knew it would be tricky to thread the needle in terms of message. So he focused on the weirdness factor. People were put off because they thought Cuccinelli was being weird. What if his messaging acknowledged that? We decided to test a message that simply stated, 'You might not agree, but at least you know where I stand.' That way, even if people thought his position was crazy, at least he could turn it into a predictable and ordered kind of crazy.

We convened focus groups, online panels and digital ad tests to try out the slogan, and it outperformed all the other messages we tried – even though it was essentially meaningless. This was a big realisation: we were able to sway voters' opinions by tailoring the candidate's message to match their psychometric tests. And because so many Republicans display these personality traits, that framing device – *I am who I am, and you know where I stand* – would probably work equally well for other Republicans. This strategy performed better among high scorers in conscientiousness who had been unsure about Cuccinelli. For them, it framed Cuccinelli as 'the devil you know' and positioned his salient 'quirkiness' as at least a reliable quirkiness.

It turns out that Republicans can accept a batshit insane candidate, *so long as it's consistent insanity*. This finding later informed almost everything that Cambridge Analytica worked on. From there, of course, it's a short jump to having a candidate brag that he could stand in the middle of Fifth Avenue and shoot somebody without losing support.

IN THE COURSE OF our experiment, we compiled reams of personal information about the people of Virginia. It was easy to get – we just bought access to it through data brokers such as Experian, Acxiom and niche firms with specialist lists from evangelical churches, media companies and so on. Even some state governments will sell you lists of hunting, fishing, or gun licensees. Did the state government bureaus care, or even bother to ask, where this data on their citizens was going? Nope. We could have been fraudsters or foreign spies and they wouldn't have had a clue.

Most people know Experian as a consumer credit reporting company. That's how it started out, calculating credit scores for people based on a variety of financial factors. The company would collect information from a wide array of sources – airline memberships, media companies, charities, even amusement parks. It also gathered information from government agencies, such as the DMV, fishing and hunting licensing and gun licensing. As it compiled these detailed profiles, the company realised it could make additional money using them for marketing.

In the 1990s, political strategists started buying personal information to use in campaigns. Think about it: if you know what kind of car or truck a person drives, whether they hunt, what charities they give to, and what magazines they subscribe to, you can start to form a picture of that person. Many Democrats and Republicans have a look. And their look was captured in this data snapshot. You can then target potential voters based on that information.

We also got access to census data. Unlike developing nations with less stringent privacy controls, the US government won't provide raw data on specific individuals, but you can get information, down to the county or neighbourhood level, on crime, obesity and illnesses such as diabetes and asthma. A census block typically contains six hundred to three thousand people, which means that by combining many sources of data, we could build models that infer those attributes about individuals. For example, by referencing risk or protective factors for diabetes, such as age, race, location, income, interest in health food, restaurant preference, gym membership and past use of weight-loss products (all of which are available in most US consumer files), we could match that data against aggregated statistics about a locality's diabetes rates. We could then create a score for each person in a given neighbourhood measuring the likelihood that they had a health issue like diabetes – even if the census or consumer file never directly provided that data on its own.

Gettleson and I spent hours exploring random and weird combinations of attributes. Were there people who had gun licenses but also belonged to the ACLU (American Civil Liberties Union)? Were there people who had season tickets to a symphony and a lifetime NRA (National Rifle Association) membership? Are gay Republicans even real? One day we found ourselves wondering whether there were donors to anti-gay churches who also shopped at organic food stores.

We did a search of the consumer data sets we had acquired for the pilot and found a handful of people whose data showed that they did both.

I instantly wanted to meet one of these mythical creatures, in part because I was curious but also because I wanted to make sure our data was accurate. We pulled the names that came up, then sent them to a call centre, where agents phoned each person to ask if they'd be willing to meet with a researcher to answer some questions. Most said no, but there was one woman who agreed – and whom I couldn't wait to meet. Her spending habits seemed all over the map – a Whole Foods shopper with an interest in yoga, but also a member of an anti-gay church and a donor to right-wing charities – which made me suspect that either our data was somehow faulty or this person was among the most fascinating characters in the United States.

The woman's data guided me to a modest split-level in the suburbs of Fairfax County. For a moment, I hesitated. 'Ugh, is this going to be awkward?' But I'd come all this way, so I walked up to the door and rang the bell. I heard wind chimes just over my head. Then a perky blonde with blown-out hair opened the door and almost leapt at me. *'HEYYYY!!! Come on in!'* As we entered the house I noticed she was indeed wearing Lululemon yoga pants. She showed me into her living room, which smelled of incense and had statues of *both* Buddha and the elephant-headed Hindu god Ganesh. Then I spotted a crucifix on the wall. It was all pretty extra.

When she offered me a glass of homemade kombucha, I accepted. In the kitchen, she opened a large jar of *something* and poured an extremely ripe and slightly coagulated liquid into a glass.

'It's really probiotic.'

'Yeah, I can tell,' I answered, looking at the floating chunks in the glass.

As we started talking, she spoke in New Age terms about trying to 'align her positive energy,' inspired, no doubt, by the Deepak Chopra on her bookshelf. But when we started talking about morality, she shifted abruptly into fire-and-brimstone evangelical views – particularly about gay people, who she *knew* were going *straight* to hell (no pun intended). Yet even the way she expressed that belief was a strange amalgamation: she said that being gay was like a block in your energy – a sinful block.

She evangelised to me for two hours as I sat there scribbling notes, as though we were participants in some kind of messed-up therapy session.

I came away from that encounter swirling with ideas. I felt like I was on to something important, because how the hell would a pollster classify this woman? This convinced me we needed to invest more in understanding the nuances behind the demographics. I once met the primatologist Jane Goodall, and she said something that always stuck with me. Mingling at a reception, I asked why she researched primates in the wild instead of in a controlled lab. It's simple, she said: because they don't live in labs. And neither do humans. If we are to *really* understand people, we have to always remember that they live outside of data sets.

It's amazing how easy it is to get drawn into something you are interested in. We were a British military contractor, working on big ideas, with a growing team of mostly gay and mostly liberal data scientists and social researchers. So why had we started working with this eclectic mix of hedge fund managers, computer scientists and a guy who ran a niche right-wing website? *Because the idea was a killer one.* With free rein to study such an abstract and fluid thing as culture, we could be breaking into a new field of researching societies. If we could put society into a computer, we could start to quantify everything and encapsulate problems like poverty and ethnic violence in a computer; we could simulate how to fix them. And just as the woman did not see the contradictions in her idols, I did not yet see the contradiction in what I was doing.

5

CAMBRIDGE ANALYTICA

IN THE COURSE OF HOME VISITS AND FOCUS GROUPS IN THE autumn of 2013, we found Virginia contained a quintessential cross section of American life. We toured from Fairfax through the middle of the state, and then headed down to Norfolk and Virginia Beach, stopping in at local bars and mom-and-pop restaurants, where we welcomed the ambience as much as the sustenance. Indeed, you can learn a lot from the ways people eat, drink and talk. Sweet tea and certain foods became pet obsessions after we discovered their cultural significance. If the American South was traditionally delineated by the old Mason-Dixon Line, which marked the border between slave owners and free states, a different divide cut through contemporary Virginia, where restaurants to the north served tea unsweetened and those to the south served it sweet. Here began the 'real South,' the locals informed us – at the Sweet Tea Line, not just south of Mason-Dixon but farther south than Richmond too.

My favourite activities were to watch and listen to the Americans who agreed to let us spend time with them. I would sit on the sofa and eavesdrop on people talking about their day, or what they heard on the radio, or office politics. I would watch people watching Fox News and notice how furious they would get (which, because I came from a country without Fox News, was one of the most interesting things to see). It was a weird performance, as they would sit down waiting – and expecting – to be insulted by whatever the 'elites' had done to them that day. They would flip on Fox and their rage became palpable. Sometimes I seemed to be witnessing a therapy session, like when people smash things in a rage room after a frustrating week. This was quite the juxtaposition to what I was normally used to seeing when my friends would happen to stumble upon Fox News. I distinctly remember Alistair Carmichael once referred to a shouty, red-faced Fox News correspondent as a 'slapped arse'.

One couple told me about the thousands of dollars they owed for their insurance deductibles and how they would sometimes skimp on their prescriptions because they had to repair their car that month. They'd agreed to the interview because the $100 they received would get them closer to covering next month's costs. But who did they blame for their insurance costs? Not their employer's bad health plan or a lack of decent pay – they blamed Obamacare. They genuinely thought it was rolled out simply to help more undocumented workers come to America in a grand plan of liberal social engineering to keep the Democrats in power through more Democrat-leaning Latino voters, which in their minds made insurance and hospitals more expensive.

People would feel better about their day after an hour-long session in the Fox News rage room – they could groan out their stress, and afterward their problems at work or home were someone else's fault. It meant that their struggles could be wholly externalised, sparing them the stark reality that maybe their employer didn't care enough about them to give them a living wage. It would be too painful to admit that perhaps they were being taken advantage of by someone they saw every day rather than the faceless enemy of Obamacare and 'illegals'.

This was my longest exposure to Fox News, and all I could think about was how the network was conditioning people's sense of identity into something that could be weaponised. Fox fuels anger with its hyperbolic narratives because anger disrupts the ability to seek, rationalise and weigh information. This leads to a psychological bias called *affect heuristic,* where people use mental shortcuts that are significantly influenced by emotion. It's the same bias that makes people say things they later regret in a fit of anger – in the heat of the moment they are, in fact, thinking differently.

With their guards down, Fox's audience is then told they are part of a group of 'ordinary Americans'. This identity is hammered home over and over, which is why there are so many references to 'us' and direct chatting to the audience by the moderators. The audience is reminded that if you are *really* an 'ordinary American,' this is how you – i.e., 'we' – think. This primes people for *identity-motivated reasoning,* which is a bias that essentially makes people accept or reject information based on how it serves to build or threaten group identity rather than on the merits of the content. This motivated reasoning is how Democrats and Republicans can watch the exact same newscast and reach the opposite conclusion. But I began to understand that Fox works because it grafts an identity onto the minds of viewers, who then begin to interpret a debate about ideas as an *attack on their identity*. This in turn triggers a *reactance effect,* whereby alternative viewpoints actually strengthen the audience's resolve in their original belief, because they sense a threat to their personal freedom. The more Democrats criticised Fox's bait, the more entrenched the audience's views and the angrier they became. This is how, for example, viewers could reject criticism of Donald Trump for saying racist things: they internalised the critique as an attack on *their own identity* rather than that of the candidate. This has an insidious effect in which the more debate occurs, the more entrenched the audience becomes.

Doing this research, I also began to see socially and economically deprived white people in a different way. It was clear that part of what underpins racist and xenophobic sentiment is a feeling of being threatened, reinforced through constant and salient 'warnings' from sources like Fox News. One of the problems I noticed with the topical political debates on American cable news channels is the lack of nuance in labelling voter constituencies. White voters, Latino voters, women voters, suburban voters, etc., are all frequently discussed as unidimensional and monolithic groups, when in fact the salient aspects of many voters' identities do not actually reflect the labels that pollsters, analysts, or consultants use to describe them. And this in turn alienates certain people. If you're a white man living in a trailer, for example, you're probably going to get angry when you are shown people on TV who are insisting that white people are super-privileged in this country. If you grew up using an outhouse, you probably don't have much tolerance for a big discussion about whether trans people should be able to use the toilets of their choice. If you're lower middle class and you see a black person on welfare, it's not surprising that your attitude would be 'Well, what about *my* welfare?' if you live in a state that has continually cut your support. This is not to defend these views, but if we want to understand them, we have to remain open to other perspectives, even ugly ones.

As part of our early exploration of American culture, we looked at two areas we thought might be at play in this social discord. First we looked at whether a sense of *social identity threat* was fuelling some of these views. The second area was related but slightly different. A common logical fallacy that people have is seeing the world as a zero-sum game of winners and losers. This flawed logic extends into a perception that attention paid to other groups will ultimately mean less attention for people like them. Either way, minorities seemed to be 'threats' – identity threats or threats to resources. Following this hypothesis of an underlying sense of threat, we wanted to see if we could mitigate some of these feelings, and we did so by trying to reduce the sense of threat. We would ask people in one study to imagine they were invincible superheroes who couldn't be harmed or killed. We then asked them about types of people they ordinarily considered threatening – gays, immigrants, people of other races – and found that they had a more muted response to those 'threatening' stimuli. If you are invincible, nothing can threaten you, not even the gays. This was fascinating for me and the team, as we were unpacking possible ways to mitigate underlying factors for racial tension. With each experiment, we learned more about how to manipulate outcomes according to people's innermost traits.

Our work in Virginia yielded promising results. We showed that there were relationships between personality traits and political outcomes, and that we could not only predict certain behaviours but also shift attitudes by framing the language of messages to correspond to psychometric profiles. We knew that even though the data sets we used were pretty decent in this little sandbox of pilots, they were still woefully inadequate for discerning all the nuances of personality and identity. In order to truly re-create society in silico, we would need to find even more complete data – way more data. But that was a problem to be solved in the future.

Nix gave us a week to write a report that would normally have taken two months to do. He was anxious to get the ball rolling, because he knew what was at stake. Bannon had told him that Mercer might invest up to $20 million. For a niche firm like SCL, which had an annual budget in the range of $7 million to $10 million, this would be a game-changing amount of money.

After pulling late nights and working through the weekend, we sent the report to Bannon the following Monday, and he immediately understood the possibilities of what we could accomplish. He was fully on board. In fact, he called the SCL office after reading the report and was almost giddy. 'This is so great, guys,' he kept saying.

Now we just had to persuade Robert Mercer.

A COUPLE OF WEEKS after this, one evening in late November 2013, Nix called me at home. 'Pack a bag,' he said. 'You're flying to New York tomorrow.' He, Tadas Jucikas and I were going to present our findings to Robert Mercer and his daughter Rebekah.

Nix flew out first thing in the morning, but for some reason he'd booked Jucikas and me on a later flight. We landed at JFK around four in the afternoon, with our meeting scheduled to start at five. As we stood in line at US customs, my phone rang. It was Nix. 'Where the fuck are you?' he demanded.

'We just got off the plane,' I told him.

'Well, you're late,' he snapped. 'You'd better hurry up and get here.'

'I can't just wave myself through passport control!' I said, exasperated. As we squabbled on the phone, others in the queue turned their heads. We continued arguing until a customs official barked at me to get off the phone. And that wasn't the end of it. Nix called me repeatedly – as we got in the car, when we arrived at the hotel, and as I changed clothes to come to the meeting. This was typical Nix, planning poorly and expecting me to fix everything. Irritated, I decided to put my phone on silent and take my time getting ready, mostly just to vex him. Jucikas and I took a cab to the meeting, which was at Rebekah Mercer's apartment on the Upper West Side. Rebekah and her husband, a French financier named Sylvain Mirochnikoff, had bought six apartments at the Heritage at Trump Place, on Riverside Boulevard, combining them into one gigantic seventeen-bedroom home. The place took up most of the twenty-third, twenty-fourth and twenty-fifth floors, with spectacular views up and down the Hudson, dotted with all the lights of New York.

But it was also tacky, as Rebekah had decorated it with random artsy-craftsy touches: ceramic figurines, throw pillows, holiday decorations. In the living room, she had a magnificent grand piano, and on top of it was a clusterfuck of knickknacks and framed family photos.

Rebekah was an interesting case. She had studied biology and mathematics at Stanford and earned a master's in operations research and engineering economic systems. She had then followed her father into trading at Renaissance Technologies but left when she began homeschooling her children. In 2006, she and her sisters bought a bakery in Manhattan, so her life became primarily about kids and chocolate chunk cookies. She had a super-perky air about her, like some kind of right-wing cheerleader. And because she had so much money to give, she was an influential person in Republican circles. Unlike more cynical Republican Party operatives, she had what Mark Block called 'TB' – she was a *true believer* in these conservative crusades.

I walked into the living room and saw Rebekah sitting on a loveseat with Nix. The two of them were chatting and laughing, Nix turning on the charm. The room was packed with people – Bob Mercer, Bannon, Block, a couple of old men from the pro-Brexit right-wing UK Independence Party (UKIP), and a collection of guys in suits who I assumed were lawyers or corporate advisers. Several other Mercers were also there, including Bob's wife, Diana, their daughter Jennifer, and a few grandchildren. This was a family affair.

Mercer was the antithesis of his daughters, who were gaudy and garrulous. He rarely looked at anyone and mostly just listened. He wore a plain grey suit, even though we were at his daughter's house for dinner. Most of the talking was done by his daughters or his entourage. He was intimidating, deeply serious and almost entirely nonverbal. When he did speak, his tone was flat. He asked only questions related to very technical aspects of our work and always wanted me to give specific statistics.

When the time came, Nix stood up and gave a short speech about SCL's pedigree, our work for the military, and how the firm didn't normally indulge private clients as we were now (a lie) but how Mercer's persistence at chasing him had worn him down. I had to stop myself from rolling my eyes. Nix then introduced me and started describing the project wholly incorrectly. He clearly had not read the long report and so just started making up findings. I knew Mercer would see right through his bullshit, so I interrupted to describe what we'd done in Virginia. Nix glared at me as he sat down beside Rebekah. In discussing the project, I added some of the more colourful details to draw the Mercers in. When I mentioned Kombucha Lady, describing her as an evangelical Christian who loved yoga and organic foods, Rebekah blurted, 'That's *so* me! Finally, someone understands us!'

I also talked about SCL's projects in other regions – the Middle East, North Africa, the Caribbean. When I got to the Trinidad project, I could see Bannon nodding his head as I described the idea of replicating society in silico. It also got Bob Mercer's attention, because as an engineer he was especially interested in this part. After I started at SCL, I had realised that the information propagation R&D projects that DARPA was funding were just cultural trend forecasting by another name. Harvesting social media data to profile users with an algorithm was just the beginning. Once their behavioural attributes were inferred, simulations could be run to map out how they would communicate and interact with one another at scale. This brought to mind experiments from the 1990s in a niche field of sociology called 'artificial societies,' which involved attempts by crude multi-agent systems to 'grow' societies in silico. I could remember as a teenager reading Isaac Asimov's *Foundation* series, where scientists used large data sets about societies to create the field of 'psychohistory,' which allowed them to not only predict the future but also control it.

Mercer had involved people from his company Renaissance Technologies in the original scoping of SCL, and, given that Nix was so focused on money and a hedge fund was part of the early stages of this project, everyone was under the impression that this was going to become a commercial venture. To put it crudely, if we could copy everyone's data profiles and replicate society in a computer – like the game *The Sims* but with real people's data – we could simulate and forecast what would happen in society and the market. This seemed to be Mercer's goal. If we created this artificial society, we thought we would be on the threshold of creating one of the most powerful market intelligence tools in the world. We would be venturing into a new field – cultural finance and trend forecasting for hedge funds.

Mercer, the computer engineer turned social engineer, wanted to re-factor society and optimise its people. One of his hobbies is building model train sets, and I got the feeling that he thought he could, in effect, get us to build him a model society for him to tinker with until it was perfect. By taking a leap at quantifying many of the intrinsic aspects of human behaviour and cultural interaction, Mercer eventually realised that he could have at his disposal the Uber of information warfare. And, like Uber, which decimated the hundred-year-old taxi industry with a single app, his venture was about to do the same with democracy.

Bannon's goal was fundamentally different. He was no traditional Republican. In fact, he hated Mitt Romney-style Republicans for what he saw as their vapid capitalism. He loathed Ayn Rand, because she objectified people into commodities. He would talk about how an economy needed a higher purpose and sometimes referred to himself as a Marxist, less out of ideology than to make a point – that Marx talked about humans fulfilling a purpose. He claimed to believe in dharma, a tenet of Hinduism and Buddhism that has to do with order in the universe and proper, harmonious ways of living. He felt his mission was to find America's purpose. In his mind, the time was right for a revolution: he saw several signals from the financial crisis and a decaying trust in institutions that foretold of a great reckoning looming on the horizon. Bannon's quest was quasi-religious, with him assuming the role of messiah.

So, like Mercer, Bannon hated 'big government,' but for his own reason – because he saw the administrative state replacing the roles played by tradition and culture. For him, the EU was a chief offender, a sterile bureaucracy replacing tradition in the extreme – leaving Europe to become an economic marketplace devoid of meaning. The western world seemed to Bannon as if it were losing its way by abandoning its cultural traditions for meaningless consumerism and a faceless state. For Bannon, this was a full-on culture war. As a self-anointed prophet, Bannon wanted a tool to peer into the future of our societies. And with what Bannon called *Facebook's God's-eye view* of each and every citizen, he could work to find the dharma for every American. In this way our research became almost spiritual for him.

Nix, Bannon and Mercer were all fascinated with Palantir, Peter Thiel's data-mining firm, whose name comes from the crystal ball, or all-seeing eye, from J. R. R. Tolkien's *Lord of the Rings*. At the time, it seemed to me that these men wanted to create their own private Palantir by investing in SCL. Imagine the possibilities for an investor like Mercer: predict the future of what people will buy and not buy, in order to make more money. If you can see a crash coming, you have the all-seeing orb for society: you might make billions overnight.

When I finished, Rebekah invited everyone into the dining room. The kitchen staff brought out plates of filet mignon with a delicate garnish, but, knowing that I didn't eat meat, Rebekah had asked the chef to prepare a special dish for me. It turned out to be grilled cheese sandwiches – I suppose at least she'd tried. She reached over to my plate to grab one and, after taking a bite, sighed with contentment. 'I actually just asked for these because I wanted one,' she confided.

'You know,' she said, 'I'm so glad that someone like you is giving us a chance. We need more of *your kind of people*.'

'Oh, what do you mean?' I asked innocently. Of course, I knew exactly what she meant, but I wanted her to say it out loud.

'The gays – who I love, by the way!'

I pondered how she squared the mental gymnastics of both loving the gays and also supporting causes to oppress them. But then again, I've been to many dinners where people talk about how much they love animals as they tear into a steak.

Rebekah wanted to entice more LGBT people into the Republican ranks, believing it would strengthen the party. She then said she loved my jacket and suggested we go shopping together sometime. Rebekah was so awkward, and so expertly manipulated by Nix, I almost felt sorry for her. But not quite.

At the end of the meal, Bob asked everyone to leave except for Nix, Rebekah and the lawyers. He had made the decision to invest – somewhere between $15 million and $20 million of his own money. 'We'll create an actual *palantír,*' Nix said. 'We'll literally be able to see what's going to happen.'

WITH UP TO $20 million in the bag, Nix was in a giddy mood. The night after our meeting, he took Jucikas and me to a lavish dinner at Eleven Madison Park, a Michelin-star venue with vaulted ceilings. He ostentatiously flipped through the wine list, then directed the waiter to bring us the Château Lafite Rothschild – a $2,000 bottle of wine.

'Get whatever you like,' he said, waving his arm grandly. This was a pleasant surprise, because, despite his wealth, Nix was cheap and complained about even the smallest expenses, such as office supplies. He once rejected an expense claim because someone bought 'too many' highlighters, saying, 'You don't need more than one.' But on this evening he ordered what seemed like dozens of dishes, an Arthurian feast. He was flush with his own magnificence.

The waiter brought the wine, and no sooner had he filled our glasses than Nix flailed his arm in conversation and knocked the bottle off the table. Hundred-dollar droplets spewed everywhere, and before the waiter even had a chance to whip off his arm towel and clean it up, Nix exclaimed, 'Bring us another!' I must have looked at him agape, because he winked and said, 'When you have $20 million, it doesn't really matter, does it?'

The night turned into a full-on bacchanalia. Somehow a couple of women in tight skirts materialised, to the evident shock of other diners. 'Chris, do you want one?' he asked me, until I reminded him that I was not into women, as if anyone needed reminding, and he blurted, 'Oh, do you want me to find you a bum boy?' I didn't know how to respond, but Nix just kept talking. He then told me a story about his time at Eton and what posh boys apparently do for fun. The whole scene was beyond mortifying, and it just kept getting worse.

At some point, the restaurant's management had to figure out what to do with us. Our bill by now was in the tens of thousands of dollars, so they couldn't just throw us out before we'd paid. Jucikas and Nix were too far gone to care, but I was sitting there watching everyone watch us. Then, in what was obviously a coordinated move, a dozen waiters suddenly fanned out in the room, whispering to the guests at the other tables. All the guests rose from their tables as one and walked into an adjoining dining room while the waiters picked up half-eaten entrees and bottles of wine and deftly resettled everyone away from the ruckus we were causing.

I've come to think that there was something darkly prescient about that night. Chaos and disruption, I later learned, are central tenets of Bannon's animating ideology. Before catalysing America's dharmic rebalancing, his movement would first need to instil chaos throughout society so that a new order could emerge. He was an avid reader of a computer scientist and armchair philosopher who goes by the name Mencius Moldbug, a hero of the alt-right who writes long-winded essays attacking democracy and virtually everything about how modern societies are ordered. Moldbug's views on 'truth' influenced Bannon and what Cambridge Analytica would become. Moldbug has written that 'nonsense is a more effective organising tool than the truth,' and Bannon embraced this. 'Anyone can believe in the truth,' Moldbug writes. 'It serves as a political uniform. And if you have a uniform, you have an army.'

Mercer's investment was used to fund an offshoot of SCL, which Bannon named Cambridge Analytica. I can only imagine what Bob and Rebekah Mercer would have thought if they had seen the pulsating hedonistic shit show their investment had enabled. Steve Bannon, on the other hand, probably would have loved it.

IT WAS SPRING 2014 and ten o'clock at night, several months after the dinner in New York, and as we sped through rural Tennessee, a sudden rush of chilly air cleared my head and lungs. The chain-smoking driver, Mark Block, had hotboxed the car, forcing us to open the windows. In the back of the car sat Gettleson and me. Clouds of nicotine escaped into the dark as we drove along desolate roads, surrounded by inky forests. I was back in the United States, setting up pilot projects for Cambridge Analytica, and Block was my guide. As SCL's introducer to Bannon, Block was excited about the potential of this project, and although he wasn't able to help build our models, he knew America like the back of his hand.

'I got some beers in the back,' Block said. 'Have one.' *Why not?* I thought, and the beer and conversation began to flow. Block was one of the more fascinating alt-right characters I'd encountered – both a super-friendly midwesterner with a warm smile and a seasoned Republican operative who had cut his teeth in the Nixon years.

'Let me tell you why Nixon was one of our best presidents,' he said out of the blue.

'Okay, I'll bite – why?'

'Because he fucked rats.'

'Wait – what?'

'Democrats. He fucked so many Democrats,' he laughs. 'Back then you could get away with anything.'

'Ohhhhh, right.'

'It's why my firm is called Block RF.'

Block had once been barred by the Wisconsin State Elections Board from running campaigns in that state because of alleged shady dealings during a judge's reelection bid, although this was later resolved through Block voluntarily paying a $15,000 fine without any admission of wrongdoing. In his years as head of the Koch brothers' Americans for Prosperity, a 501(c)(4) 'social welfare' group, he'd created a vast network of right-wing organisations that one watchdog group dubbed the 'Blocktopus.' For him, politics was not about ideas or policy – that was all bullshit for the true believers like Rebekah Mercer. For him, politics was guerrilla warfare, where he could play Che.

Another of Block's coups was Herman Cain's accidentally brilliant 'smoking ad'. As Cain's chief of staff when he ran for president in 2012, Block appeared in a campaign spot in which he mostly just rambled, the camera zooming in tight on his face. His grey moustache fell messily over his purple lips, charred from the cigarettes he smoked.

'I really believe that Herman Cain will put the 'United' back in the United States of America,' he says, shaking his head for emphasis. As the ad closes, he stares into the camera and sucks on a cigarette, casually exhaling smoke while the Krista Branch song 'I Am America!' swells in the background. This was almost shocking, as the FCC (Federal Communications Commission) had banned cigarette advertisements on TV and radio back in 1971. But it was Block's personal touch, his way of flipping the bird to political correctness.

I enjoyed hanging out with Block – he was quite an endearing guy and would always ask how you were doing, though I also knew that he wouldn't hesitate to screw you over on a campaign. The more we talked, the more it seemed to me that he didn't really believe in the hateful ideas the alt-right stood for – he just embraced the aesthetics of revolt. He relished his role as an eternal rebel with a cause inside his niche of the Republican Party, and we bonded over our mutual enjoyment of defying the establishment.

And that's how we began our work for Cambridge Analytica, the history-changing project that would fuel Brexit, the election of Donald Trump, and the death of personal privacy: hotboxing our way across the United States.

IN EARLY 2014, the first people CA sent over to the United States to do focus groups were sociologists and anthropologists, none of whom were American. This was intentional. There's a tendency among Americans to see their country as exceptional, but we wanted to study it like we would study any country, using the same language and sociological approaches. It was fascinating to explore America this way, and because I'm not American myself, I felt I was more able to cut through unquestioned assumptions of American culture and notice things that Americans don't see in themselves. When it comes to what's happening in other places, Americans will talk about 'tribes', 'regimes', 'radicalisation', 'religious extremists', 'ethnic conflicts', 'local superstitions', or 'rituals'. Anthropology is for other people, not Americans. America is supposedly this 'shining city upon a hill', a term Ronald Reagan famously popularised, adapting it from the biblical story of the Sermon on the Mount.

But when I would see evangelists prophesying the end times and woe unto the nonbelievers, when I watched a Westboro Baptist Church demonstration, when I saw a gun show with bikini-clad ladies carrying semi-automatics, when I heard white people talk about 'black thugs' and 'welfare queens', I saw a country deep in the throes of ethnic conflict, religious radicalisation and a bubbling militant insurgency. America is addicted to its own self-conception, and it wants to be exceptional. But it's not. America is just like any other country.

There were places in the United States that felt as foreign as any place I'd been. Just before the Mercers decided to invest in SCL, Nix, Jucikas and I met with possible backers in rural Virginia. A car took us from Washington, DC, through the wealthy suburbs, and then finally down a long road running deep into the woods. Eventually we came to a small clearing with a farmhouse, miles from any other sign of civilisation. The guy who was driving us said nothing, our phones no longer had a signal, and it felt like the opening scene of a horror film.

Inside the farmhouse, we were shown into a windowless boardroom with high-tech screens that came out of the ceiling. And then a group of NRA activists came in and, like clockwork, each one pulled out a gun and laid it on the table. The only time I'd seen anything like this was in Bosnia – but at least Bosnians will put their guns neatly on a rack. This was like something out of a Mafia movie, or a meeting of warlords in Afghanistan. I didn't say anything, because when a bunch of men put their guns on the table, you can't really say, 'Sorry, these guns are a little aggressive and making me uncomfortable.'

The United States has its own origin myths, its own extremist groups. At SCL, I had the displeasure of watching countless propaganda videos disseminated by ISIS and deadbeat wannabe African warlords. The way members of jihadist cults fetishise their guns is no different from the way members of the NRA fetishise theirs. I knew that if we were going to truly study America, we needed to do it as if we were studying tribal conflict – by mapping out the country's rituals, superstitions, mythologies and ethnic tensions.

Gettleson was one of the most productive researchers we sent. During the spring and summer of 2014, he went all across the United States, convening focus groups, having conversations with people, and then sending reports back to London. We would then generate theories and hypotheses to test in our quantitative research. Gettleson is an extremely charming and witty Brit, so it was easy for him to get people talking. He quickly observed a disconnect between Americans and their day-to-day politics. For example, people kept talking, unprompted, about the seemingly obscure issue of congressional term limits. They'd say over and over that the big problem in Washington is that the politicians stay in office too long and get bought by special interests. At one focus group in North Carolina, a couple of people used the phrase 'Drain the swamp', so he included that in the notes he sent back, too. CA would later study that phrase using multivariate tests on online panels of target voters, to see whether it resonated with voters.

Over a six-week period, Gettleson visited Louisiana, North Carolina, Oregon and Arkansas. In each state, Block connected him with people who would drive him around and help with logistics. I had asked him to focus on intersectionality – and in particular, finding people who were normally lumped into one category but had different political views. So he'd convene a focus group of, say, Latino Republicans, Latino Democrats and Latino independents. As in Virginia, we used a market research company to find the participants.

The results were eye-opening, even for someone who had already spent a lot of time in the United States. Gettleson's field reports, emailed from the road, described a country nearing a nervous breakdown.

In New Orleans, at a focus group of Hispanic independents, he met a hardcore conservative who declared, 'I'm not registering as a Republican because I'm a *real* conservative. I may have a Latino name, but I'm as American as they come!' At the other end of the table was a Peruvian convert to Islam who had worn her hijab to the meeting.

When the conversation shifted to guns, she told the man he might change his mind about the NRA if it were led by someone who looked like her. His answer was simple: 'I'd just go and buy another gun.' Later, the woman excused herself from the group to find a spare room where she could pray, leaving the conservative superman dumbfounded: *I don't know how to respond to this. I have a problem with it, but I can't tell a person they can't pray.*

Religion and guns were hardly the only hot-button issues Gettleson encountered in Louisiana, which was fertile ground for research, thanks to its super-varied ethnic diversity. Immigration also stoked heated debates, with more than a few almost escalating into fistfights.

A man named Lloyd, speaking with a Cajun accent that Gettleson found almost indecipherable, came across loud and clear in venting his disgust that the schools in his parish no longer taught his native French. He was furious that his granddaughter was being denied the chance to learn the 'culture and heritage' of her Cajun forebears.

It wasn't fifteen minutes before the same man launched into a rant about Latinos, how even in America they wouldn't stop speaking Spanish. Somehow, no one in the group saw the disconnect – that Lloyd could rant against Spanish people for speaking Spanish but still speak incomprehensibly in semi-French and bemoan the loss of his own heritage.

Ethnicity and race fuelled several other ugly moments. In one focus group, after hearing a chorus of complaints about President Obama, Gettleson asked, 'Does anyone *not* feel disappointed by the president?' The room was silent except for a young guy who until now had come off as exceedingly polite and courteous.

'I don't feel disappointed,' he said.

'And why is that?'

'Well, he's the first black president, so I wasn't expecting nothing.'

In that room, no one batted an eye, but other focus groups seemed to puncture partisan bubbles. All the same, full-blown arguments were not the norm; most participants made an effort to avoid conflict, even when they clearly disagreed. One exception came in Fort Smith, Arkansas, when a photo of Obama prompted a well-dressed lady to say, 'I'll go to my car and get my gun.' A younger man snapped: 'How fucking dare you! That is our president. Do not even joke about that.'

To Gettleson's eye, the woman had never in a million years considered that her views of the president might be challenged.

America's love affair with guns came up repeatedly, even in liberal bastions like Portland, Oregon, where a tattooed hipster might pause in delivering her progressive wishlist to worry aloud that the Obama administration was hell-bent on seizing her firearms. On a food run for an Oregon focus group, Gettleson watched in disbelief as the driver left his massive handgun on the driver's seat before running into

Subway to grab their sandwiches. 'I've never seen a handgun before,' Gettleson told me later. 'I'm thinking, *The car is unlocked – what if someone sees the gun, reaches in, and grabs it? Should I put it away? There seems to be a type of gun holder – should I put it in there? What if I accidentally fire it?* For two minutes of my life, I literally sat there staring at this gun as if it were a bomb in the car.'

A lot of the Oregonians Cambridge Analytica spoke to were obsessed with big government and 'Big Enviro'. One of them was the chairman of the Oregon Republican Party, Art Robinson, whose multiple losing bids for the US House of Representatives hadn't discouraged the Mercers from supporting his political ambitions. I went to visit him in his home, deep in the woods of Cave Junction, Oregon, and found him to be unhinged, even by alt-right standards.

Robinson, a biochemist who had worked with the Nobel Prize winner Linus Pauling, had two interests outside the lab: pipe organs and piss. He salvaged defunct nineteenth-century organs from churches and cathedrals all over the world and would spend hours taking them apart and reassembling them.

Robinson also collected urine from thousands of people, in an effort to discover the secrets of disease and longevity. He had become fixated on health and aging after his wife, Laurelee, died suddenly of an undetected illness at age forty-three. At the Oregon Institute of Science and Medicine, which he founded and based at his home, he ran pee through a giant spectrometer to determine its chemical composition. Animals – some dead, some alive – were everywhere. Cats, dogs, sheep and horses roamed the property, while inside, a zebra skin and the heads of a deer and a buffalo hung from the walls. Spiders had taken over the rafters and the place smelled of unwashed animals. There were several fully assembled pipe organs salvaged from old churches and cathedrals.

Robinson seemed to have tipped over the edge. He insisted that climate change was a hoax, argued that low doses of ionised radiation can be good for people, and warned that chemtrails were poisoning the population. Imagine my reaction when, a few years later, he was considered for the position of President Trump's scientific adviser.

THERE ARE TWO TYPES of billionaires: those who can never make enough money and those who, after making enough to last multiple lifetimes, turn their attention to changing the world. Mercer was the latter. Although Cambridge Analytica was created as a business, I learned later that it was never intended to make money. The firm's sole purpose was to cannibalise the Republican Party and remould American culture. When CA launched, the Democrats were far ahead of the Republicans in using data effectively. For years, they had maintained a central data system in VAN, which any Democratic campaign in the country could tap into. The Republicans had nothing comparable. CA would close that gap.

Mercer looked at winning elections as a social engineering problem. The way to 'fix society' was by creating simulations: if we could quantify society inside a computer, optimise that system, and then replicate that optimisation outside the computer, we could remake America in his image. Beyond the technology and the grander cultural strategy, investing in CA was a clever political move. At the time, I was told that because he was backing a private company rather than a PAC, Mercer wouldn't have to report his support as a political donation. He would get the best of both worlds: CA would be working to sway elections, but without any of the campaign finance restrictions that govern US elections. His giant footprints would remain hidden.

The structure chosen to set up this new entity was extremely convoluted, and it even confused staff working on projects, who were never sure who exactly they actually worked for. SCL Group would remain the 'parent' of a new US subsidiary, incorporated in Delaware, called Cambridge Analytica. For a principal investment of $15 million, Mercer took 90 per cent ownership of Cambridge Analytica, and SCL would take 10 per cent. This setup was so that CA could operate in the United States as an American company while keeping SCL's defence division a 'British' company. Therefore, SCL would not have to notify the UK Ministry of Defence or its other government clients of the new ownership and Mercer's involvement. However, this subsidiary was bestowed the IP rights to SCL's work, creating a bizarre situation where the subsidiary actually owned the core assets of its 'parent'. SCL and Cambridge Analytica then signed an exclusivity agreement whereby Cambridge Analytica would transfer all of its contracts to SCL, and SCL's personnel would service the actual delivery and work on behalf of Cambridge Analytica. And then, to allow SCL staff to use the IP that it originally gave to Cambridge Analytica, the IP was then licensed back to SCL.

Nix initially explained how this labyrinthine setup was to allow us to operate under the radar. Mercer's rivals in the finance sector watched his every move, and if they knew that he had acquired a psychological warfare firm, others in the industry might figure out his next play – to develop sophisticated trend-forecasting tools – or poach key staff. We knew Bannon wanted to work on a project with Breitbart, but this was originally supposed to be a side project to satiate his personal fixations. Of course, this was all bullshit, and they wanted to build a political arsenal. I'm not even sure Mercer knew, at first, how effective Cambridge Analytica's tools would be. He was like an investor in any startup – throwing money at smart, creative people who had an idea, in the hopes that it would turn into something valuable.

What few people know, however, is the story of who became Cambridge Analytica's very first target of disinformation. Back when Bannon and I first met, he had rejected going to a private club on Pall Mall, preferring to meet in Cambridge. Nix clocked this, realising that his normal way of courting clients – by impressing them with fancy clubs, expensive wines and fat cigars – wouldn't work on Bannon, who saw himself as an intellectual, perfectly suited to the Gothic halls and sprawling greens of Cambridge. So Nix, like some kind of mythological shape-shifter transforming to lure his prey, made an instant decision to play to that.

He told Bannon that while SCL had London offices, we were based primarily out of Cambridge because of our close partnership with the university. This was a total falsehood, made up on the fly. But for Nix, truth was whatever he deemed true in the moment. As soon as he'd said we had a Cambridge office, he started referring to it all the time, urging Bannon to stop by.

'Alexander, we don't have a Cambridge office,' I said, exasperated with his insanity. 'What the fuck are you talking about?'

'Oh, yes we do, it's just not open at the moment,' he said.

A couple of days before Bannon's next visit to the UK, Nix had the London office staff set up a fake office in Cambridge, complete with rented furniture and computers. On the day Bannon was scheduled to arrive, he said, 'Okay, everyone, we're working out of our Cambridge office today!' And we all packed up to go out there and work. Nix also hired a handful of temps and several scantily clad young women to staff the would-be office for Bannon's visit.

The whole thing felt ludicrous. Gettleson and I messaged each other, sharing links about Potemkin villages, the fake Russian towns set up in old tsarist Russia to woo Catherine the Great when she visited in 1783. We christened the office the Potemkin Site and made relentless fun of Nix for coming up with such a stupid idea. But when I walked around the fake office with Bannon, two months after I first met him in a Cambridge hotel, I could see the light in his eyes. He was buying it and loving every moment of it. Fortunately, he never noticed that some of the computers weren't actually plugged in or that some of the hired girls didn't speak English.

Nix set up the Potemkin Site every time Bannon came to town. Bannon never caught on that it was fake. Or if he did, he didn't mind. It fit the vision. And when it came time to name the new entity the Mercers were funding, Bannon chose Cambridge Analytica – because that was where we were based, he said. So Cambridge Analytica's first target was Bannon himself. The Potemkin Site perfectly encapsulated the heart and soul of Cambridge Analytica, which perfected the art of showing people what they want to see, whether real or not, to mould their behaviour – a strategy that was so effective, even a man like Steve Bannon could be fooled by someone like Alexander Nix.

6

TROJAN HORSES

'YOU KNOW DARPA FUNDS SOME OF THEIR WORK,' BRENT Clickard told me on a train ride from London to Cambridge. 'If you want to expand your team, these are the ones you want.' As one of SCL's psychologists, he floated between the company and his ongoing academic work inside one of the psychology labs at the University of Cambridge. Like me, Clickard was becoming enamoured with the possibilities of what our research could yield, which is why he was so willing to provide the firm access to the world's leading research psychologists. The psychology department at Cambridge had spearheaded several breakthroughs in using social-media data for psychological profiling, which in turn prompted interest from government research agencies. What Cambridge Analytica eventually became depended in large part on the academic research published at the university it was named after.

Cambridge Analytica was a company that took large amounts of data and used it to design and deliver targeted content capable of moving public opinion at scale. None of this is possible, though, without access to the psychological profiles of the target population – and this, it turned out, was surprisingly easy to acquire through Facebook, with Facebook's loosely supervised permissioning procedures. The story of how this came to be started in my early days at SCL, before they created Cambridge Analytica as a spin-off American brand. Brent Clickard had shown me around the Psychometrics Centre at Cambridge. Having read through many of his papers, and those of his colleagues at the centre, I was intrigued by their novel tactic of integrating machine learning with psychometric testing. It seemed like they were working on almost the same research questions we were at SCL, albeit with a slightly different purpose – or so I thought.

Research into using social data to infer the psychological disposition of individuals was published in some of psychology's top academic journals, such as the *Proceedings of the National Academy of Sciences* (*PNAS*), *Psychological Science,* and the *Journal of Personality and Social Psychology,* among many others. The evidence was clear: the patterns of a social media user's likes, status updates, groups, follows and clicks all serve as discrete clues that could accurately reveal a person's personality profile when compiled together. Facebook was frequently a supporter of this psychological research into its users and provided academic researchers with privileged access to its users' private data. In 2012, Facebook filed for a US patent for *'Determining user personality characteristics from social networking system communications and characteristics.'* Facebook's patent application explained that its interest in psychological profiling was because 'inferred personality characteristics are stored in connection with the user's profile, and may be used for targeting, ranking, selecting versions of products, and various other purposes.' So while DARPA was interested in psychological profiling for military information operations, Facebook was interested in using it for increased sales of online advertising.

As we approached the Downing Site building, I spotted a small plaque that read PSYCHOLOGICAL LABORATORY. Inside, the air was stale, and the décor hadn't been updated since at least the 1970s. We walked up a few flights of stairs and then to the last office at the end of a narrow corridor, where Clickard introduced me to Dr Aleksandr Kogan, a professor at the University of Cambridge who specialised in computational modelling of psychological traits. Kogan was a boyish-looking man and dressed as awkwardly as his manner. He stood with a simpering grin in the middle of the room, which was filled with stacks of papers and random decorations from his time studying in Hong Kong.

At first, I had no idea about Kogan's background, as he spoke

English with a perfect American accent, albeit with an exaggerated prosody. I later learned he was born in the Moldavian SSR during the final years of the Soviet Union and spent part of his childhood in Moscow. Not long after the Soviet Union collapsed, in 1991, his family emigrated to the United States, where he studied at the University of California, Berkeley, before completing his PhD in psychology in Hong Kong and joining the faculty at the University of Cambridge.

Clickard had introduced me to Kogan, as he knew the work he was doing in his lab at Cambridge could be extremely useful for SCL. But, knowing Nix's preferred style of venue, Clickard decided that the introduction should happen over canapés and wine. Nix was fickle, and he could completely write people off because he didn't like their tie or choice of restaurant. So we all met at a table booked by Clickard at an upstairs bar inside the Great Northern Hotel, beside Kings Cross train station. Kogan was visiting London for the day and had made time to tell us about his work before heading back to Cambridge. It was common enough for Nix to drink too much wine on a night out, but I'd never seen him intoxicated by a voice other than his own. The topic was social media.

'Facebook knows more about you than any other person in your life, even your wife,' Kogan told us.

Nix snapped out of his trance, reverting to his usual embarrassing self. 'Sometimes it's best wives don't know certain details,' he quipped, sipping his wine. 'Why would I ever need or want a computer to remind me – *or her*?'

'You might not want it,' the professor answered, 'but advertisers do.'

'He's interesting, but he doesn't sound like a Cambridge man to me,' mumbled Nix, drinking more wine while Kogan was in the bathroom.

'Because he's not *from* Cambridge, Alexander. *Jesus* ... He just teaches there!'

Clickard rolled his eyes. Nix was a distraction from more pressing concerns. After the firm looked at Kogan's research, Nix was eager to put him to work. SCL had just secured the financing from Mercer and was in the process of setting up a new American entity. But before Nix was to let Kogan near his new prize project in America, he would have to prove himself in the Caribbean first. At the time, in early 2014, Kogan was working with researchers based at St Petersburg State University on a psychological profiling project funded by the Russian state through a public research grant. Kogan advised a team in St Petersburg that was pulling swathes of social media profile data and using it to analyse online trolling behaviour. Given that this Russian social media research focused on maladaptive and antisocial traits, SCL thought it could be applied to the Trinidad project, as Ministry of National Security staff there were interested in experimenting with predictive modelling of Trinidadian citizens' propensity to commit crimes.

In an email to Trinidad's security ministry and its National Security Council about 'criminal psychographic profiling via [data] intercepts,' one SCL staffer said that 'we may want to either loop in or find out a bit more about the interesting work Alex Kogan has been doing for the Russians and see how / if it applies.'

Kogan eventually signed up to assist SCL on the Trinidad project, where he offered advice on how to model a set of psychological constructs that past research had identified as related to antisocial or deviant behaviour. Kogan wanted data in exchange for helping to plan the project, and he started discussions with SCL about accessing its data set of 1.3 million Trinidadians for his own research. What I liked about Kogan was that he wanted to work fast and to get stuff done, which was not common for professors accustomed to the glacial pace of academic life. And he came across as honest, ambitious and upfront, if a little bit naïve, in his excitement for ideas and intellectual ambition.

I got along quite well with Kogan in the beginning. He shared my interest in the emerging fields of computational psychology and computational sociology. We would talk for hours about the promise of behavioural simulation, and when we discussed SCL, he was palpably excited. At the same time, Kogan was slightly odd, and I noticed that his colleagues would make snide remarks about him when he wasn't around. But it wasn't as if this bothered me. If anything, it made me relate to him more – after all, I'd been on the receiving end of plenty of snide remarks myself. Besides, you had to be a bit weird to work at SCL.

When Kogan joined the Trinidad initiative in January 2014, we were just launching the early trial phases of the America project with Bannon. Based on our qualitative studies, we had some theories we wanted to test, but the available data was insufficient for psychological profiling. Consumer information – from sources like airline memberships, media companies and big-box stores – didn't produce a strong enough signal to predict the psychological attributes we were exploring. This wasn't surprising, because shopping at Walmart, for example, doesn't define who you are as a person. We could infer demographic or financial attributes, but not personality – extroverts and introverts both shop at Walmart, for example. We needed data sets that didn't just cover a large percentage of the American population but also contained data that was significantly related to psychological attributes. We suspected we needed the kind of social data we had used on other projects in other parts of the world, such as clickstreams or the types of variables observed in a census record, which Kogan had picked up on.

Kogan started on Trinidad, but he was far more intrigued by SCL's work in the United States. He told me that if he was brought on to the American job, we could work with his team at the Psychometrics Centre to fill gaps in the variables and data categories in order to create more reliable models. He started asking to access some of our data sets to see what might be missing in the training set, which is the sample data sets one uses to 'train' a model to identify patterns. But that wasn't quite the problem. Clickard told him that we'd done preliminary modelling and had training sets but that we needed data at scale. We couldn't find data sets that contained variables that we knew helped predict for psychological traits *and* covered a wide population. It was becoming a major stumbling block. Kogan said that he could solve the problem for us – as long as he could use the data for his research too. When he said that if he was brought onto the America project, we could set up the first global institute for computational social psychology at the University of Cambridge, I was instantly on board. One of the challenges for social sciences like psychology, anthropology and sociology is a relative lack of numerical data, since it's extremely hard to measure and quantify the abstract cultural or social dynamics of an entire society. That is, unless you can throw a virtual clone of everyone into a computer and observe their dynamics. It felt like we were holding the keys to unlock a new way of studying society. How could I say no to that?

In the spring of 2014, Kogan introduced me to a couple of other professors at the Psychometrics Centre. Dr David Stillwell and Dr Michal Kosinski were working with a massive data set they'd harvested legally from Facebook. They were pioneers in social-media-enabled psychological profiling. In 2007, Stillwell set up an application called myPersonality, which offered users a personality profile for joining the app. After giving the user a result, the app would harvest the profile and store it for use in research.

The professors' first paper on Facebook was published in 2012, and it quickly caught the attention of academics. After Kogan connected us, Kosinski and Stillwell told me about the huge Facebook data sets they'd acquired in their years of research. The US military's research agency, DARPA, was one of the funders of their research, they said, making them well suited to work with a military contractor. Stillwell was typically muted in our interactions, but Kosinski was clearly ambitious and tended to push Stillwell into keeping the conversation going. Kosinski knew this data could be extremely valuable, but he needed Stillwell to agree to any data transfers.

'How did you get it?' I asked.

They told me, essentially, that Facebook simply let them take it, through apps the professors had created. Facebook wants people to do research on its platform. The more it learns about its users, the more it can monetise them. It became clear when they explained how they collected data that Facebook's permissions and controls were incredibly lax. When a person used their app, Stillwell and Kosinski could receive not only that person's Facebook data, but the data of all of their friends as well. Facebook did not require express consent for apps to collect data from an app user's friends, as it viewed being a user of Facebook as enough consent to take their data – even if the friends had no idea the app was harvesting their private data. The average Facebook user has somewhere between 150 and 300 friends. My mind turned to Bannon and Mercer, as I knew they would love this idea – and Nix would simply love that they loved it.

'Let me get this straight,' I said. 'If I create a Facebook app, and a thousand people use it, I'll get ... like 150,000 profiles? *Really? Facebook actually lets you do that?*'

That's right, they said. And if a couple million people downloaded the app, then we'd get 300 million profiles, minus the overlapping mutual friends. This would be an astonishingly huge data set. Up to that point, the largest data set I had worked on was Trinidad, which I thought was quite large, with profiles of one million people. But this set was on an entirely different level. In other countries, we had to get special access to data or spend months scraping and harvesting for populations several orders of magnitude smaller.

'So how do you get people to download this app?' I asked.

'We just pay them.'

'How much?'

'A dollar. Sometimes two.'

Now, remember, I've got a potential $20 million burning a hole in our firm's pocket. And these profs have just told me that I can get tens of millions of Facebook profiles for ... a million dollars, give or take. This was a no-brainer.

I asked Stillwell if I could run some tests on their data. I wanted to see if we could replicate our results from Trinidad, where we had access to similar types of internet browsing data. If the Facebook profiles proved as valuable as I hoped, we would not only be able to fulfill Robert Mercer's desire to create a powerful tool – what was even cooler was that we could mainstream a whole new field of academia: computational psychology. We were standing at the frontier of a new science of behavioural simulation and I was bursting with excitement at the prospect.

FACEBOOK LAUNCHED IN 2004 as a platform to connect students and peers in college. In a few years, the site grew to become the largest social network in the world, a place where almost everyone – even your parents – shared photos, posted innocuous status updates and organised parties. On Facebook, you could 'like' things – pages of brands or topics, along with the posts of friends. The purpose of liking was to allow users a chance to curate their personas and follow updates from their favourite brands, bands or celebrities. Facebook considers this phenomenon of liking and sharing the basis of what it calls a 'community'. Of course, it also considers this the basis of its revenue model, where advertisers can optimise their targeting using Facebook data. The site also launched an API (application programming interface) to allow users to join apps on Facebook, which would then ingest their profile data for a 'better user experience'.

In the early 2010s, researchers quickly caught on that entire populations were organising data about themselves in one place. A Facebook page contains data on 'natural' behaviour in the home environment, minus the fingerprints of a researcher. Every scroll is tracked, every movement is tracked, every like is tracked. It's all there – nuance, interests, dislikes – and it's all quantifiable. This means the data from Facebook has increasingly more ecological validity, in that it is not prompted by a researcher's questions, which inevitably inject some kind of bias. In other words, many of the benefits of the passive qualitative observation traditionally used in anthropology or sociology could be maintained, but as many social and cultural interactions were now captured in digital data, we could add the benefits of generalisability one achieves in quantitative research. Previously, the only way one could have acquired such data would have been from your bank or phone company, which are strictly regulated to prevent access to that sort of private information. But unlike a bank or telecom company, social media operated with virtually no laws governing its access to extremely granular personal data.

Although many users tend to distinguish between what happens online from what happens *IRL* (in real life), the data that is generated from their use of social media – from posting reactions to the season finale of a show to liking photos from Saturday night out – is generated from life outside the internet. In other words, Facebook data *is* IRL data. And it is only increasing as people live their lives more and more on their phones and on the internet. This means that, for an analyst, there's often no need to ask questions: you simply create algorithms that find discrete patterns in a user's naturally occurring data. And once you do that, the system itself can reveal patterns in the data that you otherwise would have never noticed.

Facebook users curate themselves all in one place in a single data form. We don't need to connect a million data sets; we don't have to do complicated maths to fill in missing data. The information is already in place, because everyone serves up their real-time autobiography, right there on the site. If you were creating a system from scratch to watch and study people, you couldn't do much better than Facebook.

In fact, a 2015 study by Youyou, Kosinski and Stillwell showed that, using Facebook likes, a computer model reigned supreme in predicting human behaviour. With ten likes, the model predicted a person's behaviour more accurately than one of their co-workers. With 150 likes, better than a family member. And with 300 likes, the model knew the person better than *their own spouse*. This is in part because friends, colleagues, spouses and parents typically see only part of your life, where your behaviour is moderated by the context of that relationship. Your parents may never see how wild you can get at a 3 a.m. rave after dropping two hits of MDMA, and your friends may never see how reserved and deferential you are in the office with your boss. They all have slightly different impressions of who you are. But Facebook peers into your relationships, follows you around in your phone, and tracks what you click and buy on the internet. This is how data from the site becomes more reflective of who you 'really are' than the judgments of friends or family. In some respects, a computer model can know a person's habits better than they even know themselves – a finding that compelled the researchers to add a warning. 'Computers outpacing humans in personality judgment,' they wrote, 'presents significant opportunities and challenges in the areas of psychological assessment, marketing and privacy.'

With access to enough Facebook data, it would finally be possible to take the first stab at simulating society *in silico*. The implications were astonishing: you could, in theory, simulate a future society to create problems like ethnic tension or wealth disparity and watch how they play out. You could then backtrack and change inputs, to figure out how to mitigate those problems. In other words, you could actually start to model solutions to real-world issues, but inside a computer. For me, this whole idea of *society as a game* was super epic. I was obsessed with the idea of the institute that Kogan suggested to me, and became extremely eager to somehow make it happen. And it wasn't just our pet obsession; professors all over were getting just as enthused. After meetings at Harvard, Kogan emailed me about their feedback, saying, 'The operative term is game changing and revolutionising social science.' And at first, it seemed like Stillwell and Kosinski were excited, too. Then Kogan let slip to them that CA had a budget of $20 million. And all the academic camaraderie ground immediately to a halt.

Kosinski sent Kogan an email saying they wanted half a million dollars up front, plus 50 per cent of all 'royalties' for the use of their Facebook data. We had not even proven this could work at scale in a field trial yet, and they were already demanding huge amounts of money. Nix told me to refuse, and this made Kogan panic that the project was going to fall apart before it even began. So the day after we rejected Kosinski's demand for cash, Kogan said he could do it on his own, on his original terms – he would help us get the data, CA would pay for it at cost, and he would get to use it for his research. Kogan said he had access to more apps that had the same friends-collection permission from Facebook and that he could use those apps. I was immediately wary, thinking that Kogan might have just been planning to use Stillwell and Kosinski's app under the radar. But Kogan insisted to me that he'd built his own. 'Okay,' I said. 'Prove it. Give me a dump of data.' To make sure these were not just pulled from the other app, we gave Kogan $10,000 to pilot his new app with a new data set. He agreed and did not ask for any money for himself, so long as he could keep a copy of the data.

Although he never told me this at the time, Kosinski has since said that he intended to give the money from the licensing of the Facebook data to the University of Cambridge. However, the University of Cambridge also strongly denies that it was involved with any Facebook data projects, so it is unclear that the university was aware of this potential financial arrangement, or would have accepted the funds if offered.

The following week, Kogan sent SCL tens of thousands of Facebook profiles, and we did some tests to make sure the data was as valuable as we'd hoped. It was even better. It contained complete profiles of tens of thousands of users – name, gender, age, location, status updates, likes, friends – *everything*. Kogan said his Facebook app could even pull private messages. 'Okay,' I told him. 'Let's go.'

WHEN I STARTED WORKING with Kogan, we were eager to set up an institute that would warehouse the Facebook, clickstream and consumer data we were collecting for use by psychologists, anthropologists, sociologists, data scientists – any academics who were interested. Much to the delight of my fashion professors at UAL, Kogan even let me add several clothing-style and aesthetic items that I could test for my PhD research. We planned to go to different universities around the world, continuing to build up the data set so we could then start modelling things in the social sciences. After some professors at Harvard Medical School suggested we could access millions of their patient genetic profiles, even I was surprised at how this idea was evolving. Imagine the power, Kogan told me, of a database that linked up a person's live digital behaviour with a database of their *genes*. Kogan was excited – with genetic data, we could run powerful experiments unpacking the nature-*vs*-nurture debate. We knew we were on the cusp of something big.

We got our first batch of data through a micro-task site called Amazon MTurk. Originally, Amazon built MTurk as an internal tool to support an image-recognition project. Because the company needed to train algorithms to recognise photographs, the first step was to have humans label them manually, so the AI would have a set of correctly identified photos to learn from. Amazon offered to pay a penny for each label, and thousands of people signed up to do the work.

Seeing a business opportunity, Amazon spun out MTurk as a product in 2005, calling it '*artificial artificial intelligence*'. Now other companies could pay to access people who, in their spare time, were willing to do micro-tasks – such as typing in scans of receipts or identifying photographs – for small amounts of money. It was humans doing the work of machines, and even the name MTurk played on this. MTurk was short for 'Mechanical Turk,' an eighteenth-century chess-playing 'machine' that had amazed crowds but was actually a small man hiding in a box, manipulating the chess pieces through specially constructed levers.

Psychologists and university researchers soon discovered that MTurk was a great way to leverage large numbers of people to fill out personality tests. Rather than have to scrounge for undergraduates willing to take surveys, which never gave a truly representative sample anyway, researchers could draw all kinds of people from all over the world. They would invite MTurk members to take a one-minute test, paying them a small fee to do so. At the end of the session, there would be a payment code, which the person could input on their Amazon page, and Amazon would transfer payment into the person's account.

Kogan's app worked in concert with MTurk: a person would agree to take a test in exchange for a small payment. But in order to get paid, they would have to download Kogan's app on Facebook and input a special code. The app, in turn, would take all the responses from the survey and put those into one table. It would then pull all of the user's Facebook data and put it into a second table. And then it would pull all the data for all the person's Facebook friends and put that into another table.

Users would fill out a wide battery of psychometric inventories, but it always started with a peer-reviewed and internationally validated personality measure called the IPIP NEO-PI, which presented hundreds of items, like 'I keep others at a distance,' 'I enjoy hearing new ideas,' and 'I act without thinking'. When these responses were combined with Facebook likes, reliable inferences could then be made. For example, extroverts were more likely to like electronic music and people scoring higher in openness were more likely to like fantasy films, whereas more neurotic people would like pages such as 'I hate it when my parents look at my phone.' But it wasn't simply personality traits we could infer. Perhaps not surprisingly, American men on Facebook who liked Britney Spears, MAC Cosmetics or Lady Gaga were slightly more likely to be gay. Although each like taken in isolation was almost always too weak to predict anything on its own, when those likes were combined with hundreds of other likes, as well as other voter and consumer data, then powerful predictions could be made. Once the profiling algorithm was trained and validated, it would then be turned onto the database of Facebook friends. Although we did not have surveys for the friend profiles, we had access to their likes page, which meant that the algorithm could ingest the data and infer how they likely would have responded to each question if they had taken a survey.

As the project grew over the summer, more constructs were explored, and Kogan's suggestions began to match exactly what Bannon wanted. Kogan outlined that we should begin examining people's life satisfaction, fair-mindedness (fair or suspicious of others), and a construct called 'sensational and extreme interests,' which has been used increasingly in forensic psychology to understand deviant behaviour. This included 'militarism' (guns and shooting, martial arts, crossbows, knives), 'violent occultism' (drugs, black magic, paganism), 'intellectual activities' (singing and making music, foreign travel, the environment), 'occult credulousness' (the paranormal, flying saucers), and 'wholesome interests' (camping, gardening, hiking). My personal favourite was a five-point scale for 'belief in star signs,' which several of the gays in the office joked we should spin off into an 'astrological compatibility' feature and link it to the gay dating app Grindr.

Using Kogan's app, we would not only get a training set that gave us the ability to create a really good algorithm – because the data was so rich, dense and meaningful – but we also got the extra benefit of hundreds of additional friend profiles. All for $1 to $2 per app install. We finished the first round of harvesting with money left over. In management, they always say there is a golden rule for running any project: you can get a project done *cheap, fast, or well*. But the catch is you can choose only two, because you'll never get all three. For the first time in my life, I saw that rule totally broken – because the Facebook app Kogan created was faster, better and cheaper than anything I could have imagined.

THE LAUNCH WAS PLANNED for June 2014. I remember it was hot: even though the summer was coming, Nix kept the air-conditioning off to lower the office bills. We had spent several weeks calibrating everything, making sure the app worked, that it would pull in the right data, and that everything matched when it injected the data into the internal databases. One person's response would, on average, produce the records of three hundred other people. Each of those people would have, say, a couple hundred likes that we could analyse. We needed to organise and track all of those likes. How many possible items, photos, links and pages are there to like across all of Facebook? Trillions. A Facebook page for some random band in Oklahoma, for example, might have twenty-eight likes in the whole country, but it still counts as its own like in the feature set. A lot of things can go wrong with a project of such size and complexity, so we spent a lot of time testing the best way to process the data set for when it scaled. Once we were confident that everything worked, it was time to launch the project. We put $100,000 into the account to start recruiting people via MTurk, then waited.

We were standing by the computer, and Kogan was in Cambridge. Kogan launched the app, and someone said, 'Yay'. With that, we were live.

At first, it was the most anticlimactic project launch in history. Nothing happened. Five, ten, fifteen minutes went by, and people started shuffling around in anticipation. 'What the fuck is this?' Nix barked. 'Why are we standing here?' But I knew that it would take a bit of time for people to see the survey on MTurk, fill it out, then install the app to get paid. Not long after Nix started complaining, we saw our first hit.

Then the flood came. We got our first record, then two, then twenty, then a hundred, then a thousand – all within seconds. Jucikas added a random beeping sound to a record counter, mostly because he knew Nix had a thing for stupid sound effects, and he found it amusing how easy it was to impress Nix with gimmicky tech clichés. Jucikas's computer started going *boop-boop-boop* as the numbers went insane. The increments of zeroes just kept building, growing the tables at exponential rates as friend profiles were added to the database. This was exciting for everyone, but for the data scientists among us, it was like an injection of pure adrenaline.

Jucikas, our suave chief technology officer, grabbed a bottle of champagne. He was always full of bonhomie, the life of the party, and he made sure we had a case of champagne in the office at all times for just such occasions. He had grown up extremely poor on a farm in the waning days of the Lithuanian SSR, and over the years he had remade himself into a Cambridge elite, a dandy whose motto seemed to be live it up today, because tomorrow you might die. With Jucikas, everything was extreme and over the top. That's why he'd bought for the office an antique sabre from the Napoleonic Wars, which he now intended to use. Why open champagne the normal way when you can use a sabre?

He grabbed a bottle of Perrier-Jouët Belle Epoque (his favourite), loosened the cage holding the cork, held the bottle at an angle, and elegantly swiped the sabre down the side. The entire top snapped clean off, and champagne gushed out. We filled the flutes and toasted our success, enjoying the first of many bottles we would drink that night. Jucikas went on to explain that sabring champagne is not about brute force; it's about studying the bottle and hitting the weakest spot with graceful precision. Done correctly, this requires very little pressure – you essentially let the bottle break itself. You hack the bottle's design flaw.

WHEN MERCER FIRST MADE the investment, we assumed we had a couple of years to get the project fully running. But Bannon shot that notion down right away. 'Have it ready by September,' he said. When I suggested that was too quick, he said, 'I don't care. We just gave you millions, and that's your deadline. Figure it out.' The 2014 midterms were coming, and he wanted what he now started referring to as Project Ripon – named after the small town in Wisconsin where the Republican Party was formed – to be up and running. Many of us rolled our eyes at Bannon, who started to get weirder and weirder after the investment. But we thought we just had to placate his niche political obsessions to achieve our potential at creating something revolutionary in science. The ends would justify the means, we kept telling ourselves.

He started travelling to London more frequently, to check on our progress. One of those visits happened to be not long after we launched the app. We all went into the boardroom again, with the giant screen at the front of the room. Jucikas made a brief presentation before turning to Bannon.

'Give me a name.'

Bannon looked bemused and gave a name.

'Okay. Now give me a state.'

'I don't know,' he said. 'Nebraska.'

Jucikas typed in a query, and a list of links popped up. He clicked on one of the many people who went by that name in Nebraska – and there was everything about her, right up on the screen. Here's her photo, here's where she works, here's her house. Here are her kids, this is where they go to school, this is the car she drives. She voted for Mitt Romney in 2012, she loves Katy Perry, she drives an Audi, she's a bit basic ... and on and on and on. We knew everything about her – and for many records, the information was updated in real time, so if she posted to Facebook, we could see it happening.

And not only did we have all her Facebook data, but we were merging it with all the commercial and state bureau data we'd bought as well. And imputations made from the US Census. We had data about her mortgage applications, we knew how much money she made, whether she owned a gun. We had information from her airline mileage programs, so we knew how often she flew. We could see if she was married (she wasn't). We had a sense of her physical health. And we had a satellite photo of her house, easily obtained from Google Earth. We had re-created her life in our computer. She had no idea.

'Give me another,' said Jucikas. And he did it again. And again. And by the third profile, Nix – who'd hardly been paying attention at all – suddenly sat up very straight.

'Wait,' he said, his eyes widening behind his black-rimmed glasses. 'How many of these do we have?'

'What the fuck?' Bannon interjected with a look of annoyance at Nix's disengagement with the project.

'We're in the tens of millions now,' said Jucikas. 'At this pace, we could get to 200 million by the end of the year with enough funding.'

'And we know literally everything about these people?' asked Nix.

'Yes,' I told him. 'That's the whole point.'

The light went on: this was the first time Nix truly understood what we were doing. He could not have been less interested in things like 'data' and 'algorithms,' but seeing actual people onscreen, knowing everything about them, had seized his imagination.

'Do we have their phone numbers?' Nix asked. I told him we did. And then, in one of those moments of weird brilliance he occasionally had, he reached for the speakerphone and asked for the number. As Jucikas relayed it to him, he punched in the number.

After a couple of rings, someone picked up. We heard a woman say 'Hello?' and Nix, in his most posh accent, said, 'Hello, ma'am. I'm terribly sorry to bother you, but I'm calling from the University of Cambridge. We are conducting a survey. Might I speak with Ms Jenny Smith, please?' The woman confirmed that she was Jenny, and Nix started asking her questions based on what we knew from her data.

'Ms Smith, I'd like to know, what is your opinion of the television show *Game of Thrones*?' Jenny raved about it – just as she had on Facebook. 'Did you vote for Mitt Romney in the last election?' Jenny confirmed that she had. Nix asked whether her kids went to such-and-such elementary school, and Jenny confirmed that, too. When I looked over at Bannon, he had a huge grin on his face.

After Nix hung up with Jenny, Bannon said, 'Let me do one!' We went around the room, all of us taking a turn. It was surreal to think that these people were sitting in their kitchen in Iowa or Oklahoma or Indiana, talking to a bunch of guys in London who were looking at satellite pictures of where they lived, family photos, all of their personal information. Looking back, it's crazy to think that Bannon – who then was a total unknown, still more than a year away from gaining infamy as an adviser to Donald Trump – sat in our office calling random Americans to ask them personal questions. And people were more than happy to answer him.

We had done it. We had reconstructed tens of millions of Americans *in silico,* with potentially hundreds of millions more to come. This was an epic moment. I was proud that we had created something so powerful. I felt sure it was something that people would be talking about for decades.

7

THE DARK TRIAD

BY AUGUST 2014, JUST TWO MONTHS AFTER WE LAUNCHED THE app, Cambridge Analytica had collected the complete Facebook accounts of more than 87 million users, mostly from America. They soon exhausted the list of MTurk users and had to engage another company, Qualtrics, a survey platform based in Utah. Almost immediately, CA became one of their top clients and started receiving bags of Qualtrics-branded goodies. Jucikas would sashay around wearing an I ❤ QUALTRICS T-shirt under his otherwise perfectly tailored Savile Row suit, which everyone found both amusing and ridiculous. CA would get invoices sent from Provo, billing them each time for twenty thousand new users in their 'Facebook Data Harvest Project'.

As soon as CA started collecting this Facebook data, executives from Palantir started making inquiries. Their interest was apparently piqued when they found out how much data the team was gathering – and that Facebook was just letting CA do it. The executives CA met with wanted to know how the project worked, and soon they approached our team about getting access to the data themselves.

Palantir was still doing work for the NSA and GCHQ. Staffers there told CA that working with Cambridge Analytica could potentially open an interesting legal loophole. At a meeting in the summer of 2014 at Palantir's UK head office, in Soho Square, it was pointed out that government security agencies, along with contractors like Palantir, couldn't legally mass-harvest personal data on American citizens, but – here's the catch – polling companies, social networks and private companies could. And despite the ban on directly surveilling Americans, I was told that US intelligence agencies were nonetheless able to make use of information on American citizens that was 'freely volunteered' by US individuals or companies. After hearing this, Nix leaned in and said, 'So you mean American polling companies ... *like us*.' He grinned. I didn't think anyone was actually being serious, but I soon realised that I underestimated everyone's interest in accessing this data.

Some of the staff working at Palantir realised that Facebook had the potential to become the best discreet surveillance tool imaginable for the NSA – that is, if that data was 'freely volunteered' by another entity. To be clear, these conversations were speculative, and it is unclear if Palantir itself was actually aware of the particulars of these discussions, or if the company received any CA data. The staff suggested to Nix that if Cambridge Analytica gave them access to the harvested data, they could then, at least in theory, legally pass it along to the NSA. In this vein, Nix told me we urgently needed to make an arrangement with staff at Palantir happen, 'for the defence of our democracy'. But that, of course, was not why Nix gave them full access to the private data of hundreds of millions of American citizens. Nix's dream, as he had confided in our very first meeting, was to become the 'Palantir of propaganda'.

One lead data scientist from Palantir began making regular trips to the Cambridge Analytica office to work with the data science team on building profiling models. He was occasionally accompanied by colleagues, but the entire arrangement was kept secret from the rest of the CA teams – and perhaps Palantir itself. I can't speculate about why, but the Palantir staff received Cambridge Analytica database logins and emails with fairly obvious pseudonyms like 'Dr Freddie Mac' (after the mortgage company that was bailed out by the federal government in the 2008 housing crisis). I do know that after Palantir data scientists started building their own Facebook harvesting apps and scrapers, Nix asked them to stay after hours to keep working on applications that could replicate the Facebook data Kogan was getting – without the need for Kogan. It was no longer simply Facebook apps that were being used. Cambridge Analytica began testing innocuous-looking browser extensions, such as calculators and calendars, that pulled access to the user's Facebook session cookies, which in turn allowed the company to log in to Facebook as the target user to harvest their data and that of their friends. These extensions were all submitted – and approved – in the independent review processes of several popular web browsers.

It wasn't clear whether these Palantir executives were visiting CA officially or 'unofficially', and Palantir has since asserted that it was only a single staff member who worked at CA in a 'personal capacity'. I honestly didn't know who or what to believe at this point. As he often did with contractors on projects in Africa, Nix would bring to the office bags filled with US currency and pay contractors in cash. As contractors would work, Nix would sit at his desk flicking through the green bills, counting them into small piles, each worth thousands of dollars. Sometimes contractors were given tens of thousands of dollars each week.

Many years before, Nix had been rejected by Britain's foreign intelligence service, MI6. He often joked about it, saying it had happened because he wasn't boring enough to blend into a crowd, but the rejection had obviously stung him. Now he almost didn't care who got access to CA's data; he would have shown it to anyone, just to hear how amazing he was.

BY LATE SPRING 2014, Mercer's investment had spurred a hiring spree of psychologists, data scientists and researchers. Nix brought on a new team of managers to organise the fast-growing research operations. Although I remained the titular director of research, the new operations managers were now given control over direct oversight and planning of this rapidly growing exercise. New projects seemed to pop up each day, and sometimes it was unclear how or why projects were being approved to go to field. I complained to Nix that I was losing track of who was doing what, but he didn't see the problem. Nix simply couldn't see beyond prestige and money. He told me that most people would be grateful to be given less responsibility and work to do but still allowed to keep their title.

At this point, I did start to feel weird about everything, but whenever I spoke with other people at the firm, we all managed to calm one another down and rationalise everything. Nix would talk about shady things, but that's just who he was and no one took him seriously. And after Mercer installed Bannon, I overlooked or explained away things that, in hindsight, were obvious red flags. Bannon had his 'niche' political interests, but Mercer seemed to be too serious a character to dabble in Bannon's trashy political sideshows. The potential for our work to benefit Mercer's financial interests made so much more sense as an explanation for why he would spend all this money on something so highly speculative. Mercer literally gave CA tens of millions of dollars before the firm had acquired any data or built any software in America. From any investor's perspective, this would have been a high-risk seed capital investment. But CA also knew Mercer was not dumb or reckless, and he would have calculated the risk carefully. At the time, many on the team simply assumed that to justify taking such a high financial risk on our ideas, Mercer must have expected that the research had the chance of making tons of money at his hedge fund. In other words, the firm was not there to build an alt-right insurgency, it was there to help Mercer make money, and Nix's conspicuous love of money reinforced everyone's assumptions.

Of course, we know now that none of that happened. I don't know what else to say other than I was more naïve than I thought I was at the time. Even though I had a great deal of experience for my age, I was only twenty-four and clearly still had a lot of learning to do. When I joined SCL, I was there to help the firm explore areas like counter-radicalisation in order to help Britain, America and their allies defend themselves against new threats emerging online. I began to get accustomed to the unusual environment of this line of work, which normalised a lot of things that would seem weird to a casual observer. Information operations is not your average nine-to-five desk job, and the people or situations you encounter are all a bit odd. And anytime someone would ask about the ethics of a surreptitious project in a far-off country, they would be mocked for their naïveté about how the rest of the world 'really worked'.

It was the first time I was allowed to explore ideas without the constraints of petty internal politics or people snubbing an idea just because it had never been tried before. As much as Nix was a dick, he did give me a lot of leeway to try out new ideas. After Kogan joined, I had professors at the University of Cambridge constantly fawning over the groundbreaking potential that the project could have for advancing psychology and sociology, which made me feel like I was on a mission. And if their colleagues at universities like Harvard or Stanford were also getting interested in our work, I thought that surely we must be onto something. The institute that Kogan proposed really was an inspiring idea to me, and I saw how unlocking this data for researchers around the world could contribute so much value to so many fields. As corny as this might sound, it really felt like I was working on something important – not just for Mercer or the company, but for science. However, I let this feeling distract me to the point of allowing myself to excuse the inexcusable. I told myself that truly learning about society includes delving into uncomfortable questions about our darker sides. How could we understand racial bias, authoritarianism or misogyny if we did not explore them? What I did not appreciate is the fine line between exploring something and actually creating it.

Bannon had assumed control of the company, and he was an ambitious and surprisingly sophisticated cultural warrior. He felt that the identity politics of Democrats, with their focus on racial or ethnic blocs of voters, was actually less powerful than that of Republicans, who often insisted that American identity went beyond skin colour, religious preference or gender. A white man living in a trailer park doesn't see himself as a member of a privileged class, though others may see him that way just because he's white. Every mind contains multitudes. And Bannon's new job was to figure out how to target people accordingly.

I told Bannon that the most striking thing CA had noticed was how many Americans felt closeted – and not just gay people. This first came up in focus groups and later was confirmed in quantitative research done via online panels. Straight white men, particularly ones who were older, had grown up with a value set that granted them certain social privileges. Straight white men did not have to moderate their speech around women or people of colour, because casual racism and misogyny were normalised behaviours. As social norms in America evolved, these privileges began to erode and many of these men were experiencing challenges to their behaviour for the first time.

At the workplace, 'casual flirting' with female secretaries now imperiled your job, and talking about the 'thugs' in the African American part of town could get you shunned by peers. These encounters were often uncomfortable and threatening to their identity as 'regular men'.

Men who were not used to moderating their impulses, body language and speech began to resent what they saw as the unfair mental and emotional labour it took to change and constantly correct how they presented in public. What I found interesting was how similar the discourse that emerged from these groups of angry straight men was to liberation discourse from gay communities. These men began to experience the burden of the closet, and they did not like the feeling of having to change who they felt they were in order to 'pass' in society. Although there were very different reasons for the closeting of gays and the closeting of racists and misogynists, these straight white men nonetheless felt a subjective experience of oppression in their own minds. And they were ready to emerge from the closet and return to a time when America was great – for them.

'Think about it,' I said to Bannon. 'The message at a Tea Party rally is the same as at a Gay Pride parade: *Don't tread on me! Let me be who I am!*' Embittered conservatives felt like they couldn't be 'real men' anymore, because women wouldn't date men who behaved the way men had behaved for millennia. They had to hide their true selves to please society – and they were pissed about it. In their minds, feminism had locked 'real men' in the closet. It was humiliating, and Bannon knew that there was no force more powerful than a humiliated man. It was a state of mind he was eager to explore (and exploit).

The incel community, just coming to the fore when Cambridge Analytica was being established, was the kind of group he had in mind. Incels, or 'involuntary celibates', were men who felt ignored and chastised by a society – particularly women – that did not value average men anymore. An offshoot of the Men's Rights Movement, the incel community was in part propelled by the increasing economic inequality depriving young millennial men from accessing the same kinds of well-paying jobs their fathers had. This economic deprivation was coupled with increasingly unattainable body image standards for men in conventional and social media (without the same public recognition of male body issues or gendered pressures as for women) and the growing importance placed on physical looks in a dating scene increasingly defined by swiping left or right on a split-second glance at a photo. And as women had become more economically independent, they could afford to be more selective about their partners. Deprived of good looks and a respectable paycheck, 'average men' faced a hard reality of constant romantic rejection.

Some of these men began congregating on forums like 4chan, which grew into a repository of memes, weird fantasy fandoms, niche porn, pop culture and the countercultural reactions of frustrated youth in an increasingly atomised society. In the early 2010s, nihilistic discussions began among young men who were resigned to lives of loneliness. A new vocabulary emerged to describe their circumstances, including 'betas' (inferior men), 'alphas' (superior men), 'vocels' (voluntary celibates), MGTOW (Men Going Their Own Way, walking away from women), 'incels' (involuntary celibates), and 'robots' (incels with Asperger's).

Irrespective of the privileges afforded to them as straight white men, these groups lacked identity, direction and a sense of self-worth and grabbed on to anything that instilled a feeling of belonging and solidarity. Self-defining as the 'beta' males of society, many incels would talk about accepting the 'black pill' – a moment of reckoning with what they believed were certain innate truths about sexual and romantic attraction. Forums would include topics such as 'suicide fuel', which were examples from their daily lives of rejection that reinforced their feelings of hopelessness and ugliness. For many incels, this angry desperation had morphed into extreme misogyny.

The doctrine of the black pill was bleak and rigid, stating that only physical looks matter to women, and that certain features, including race, fall into a hierarchy of sexual desirability. Incels would share graphs and observations signalling an innate advantage for white men, as women from all races would accept a white partner, and a strong disadvantage for Asian men. To be fat or poor or old or disabled or a person of colour was to be a member of America's most unwanted. Nonwhite incels would use terms like 'JBW' – 'just be white' – as a way of trying to explain or mitigate what they saw as their innate racial disadvantages. There was a surprising amount of open recognition of white privilege, but incel discourse would frame this privilege as part of the inherent racial superiority of white men, at least in the context of sexual selection.

Ongoing jokes and memes would be shared about resisting their life sentences and waging a Beta Rebellion or Beta Uprising to fight for the redistribution of sex for the betas. But lurking behind the strange humour was the rage of a life of rejection. In scrolling through these narratives of victimhood, my mind turned back to the narratives of extreme jihadist recruitment media, with the same naïve romanticism of oppressed men breaking the shackles of a vapid society to transform themselves into glorified heroes of rebellion. Likewise, these incels were perversely attracted to society's 'winners,' like Donald Trump and Milo Yiannopoulos, who in their warped view represented the epitome of the same hypercompetitive alphas who brutalised them, to lead the charge. Many of these seething young men were ready to burn society to the ground. Bannon sought to give them an outlet via Breitbart, but his ambition didn't stop there. He saw these young men as the early recruits in his future insurgency.

When Cambridge Analytica launched, in the summer of 2014, Bannon's goal was to change politics by changing culture; Facebook data, algorithms and narratives were his weapons. First we used focus groups and qualitative observation to unpack the perceptions of a given population and learn what people cared about – term limits, the deep state, draining the swamp, guns and the concept of walls to keep out immigrants were all explored in 2014, several years before the Trump campaign. We then came up with hypotheses for how to sway opinions. CA tested these hypotheses with target segments in online panels or experiments to see whether they performed as the team expected, based on the data. We also pulled Facebook profiles, looking for patterns in order to build a neural network algorithm that would help us make predictions.

A select minority of people exhibit traits of narcissism (extreme self-centredness), Machiavellianism (ruthless self-interest), and psychopathy (emotional detachment). In contrast to the Big Five traits found in everyone to some degree as part of normal psychology – openness, conscientiousness, extroversion, agreeableness and neuroticism – these 'dark triad' traits are maladaptive, meaning that those who exhibit them are generally more prone to antisocial behaviour, including criminal acts. From the data CA collected, the team was able to identify people who exhibited neuroticism and dark-triad traits, and those who were more prone to impulsive anger or conspiratorial thinking than average citizens. Cambridge Analytica would target them, introducing narratives via Facebook groups, ads or articles that the firm knew from internal testing were likely to inflame the very narrow segments of people with these traits. CA wanted to provoke people, to get them to engage.

Cambridge Analytica did this because of a specific feature of Facebook's algorithm at the time. When someone follows pages of generic brands like Walmart or some prime-time sitcom, nothing much changes in his newsfeed. But liking an extreme group, such as the Proud Boys or the Incel Liberation Army, marks the user as distinct from others in such a way that a recommendation engine will prioritise these topics for personalisation. Which means the site's algorithm will start to funnel the user similar stories and pages – all to increase engagement. For Facebook, rising engagement is the only metric that matters, as more engagement means more screen time to be exposed to advertisements.

This is the darker side of Silicon Valley's much celebrated metric of 'user engagement'. By focusing so heavily on greater engagement, social media tends to parasitise our brain's adaptive mechanisms. As it happens, the most engaging content on social media is often horrible or enraging. According to evolutionary psychologists, in order to survive in premodern times, humans developed a disproportionate attentiveness toward potential threats. The reason we instinctually pay more attention to the blood and gore of a rotting corpse on the ground than to marvelling at the beautiful sky above is that the former was what helped us survive. In other words, we evolved to pay keen attention to potential threats. There's a good reason you can't turn away from grisly videos: *you're human*.

Social media platforms also use designs that activate 'ludic loops' and 'variable reinforcement schedules' in our brains. These are patterns of frequent but irregular rewards that create anticipation, but where the end reward is too unpredictable and fleeting to plan around. This establishes a self-reinforcing cycle of uncertainty, anticipation and feedback. The randomness of a slot machine prevents the player from being able to strategise or plan, so the only way to get a reward is to keep playing. The rewards are designed to be just frequent enough to reengage you after a losing streak and keep you going. In gambling, a casino makes money from the number of turns a player takes. On social media, a platform makes money from the number of clicks a user performs. This is why there are infinite scrolls on newsfeeds – there is very little difference between a user endlessly swiping for more content and a gambler pulling the slot machine lever over and over.

IN THE SUMMER OF 2014, Cambridge Analytica began developing fake pages on Facebook and other platforms that looked like real forums, groups and news sources. This was an extremely common tactic that Cambridge Analytica's parent firm SCL had used throughout its counterinsurgency operations in other parts of the world. It is unclear who inside the firm actually gave the final order to set up these disinformation operations, but for many of the old guard who had spent years working on projects around the world, none of this seemed unusual. They were simply treating the American population in the exact same way they would treat the Pakistani or Yemeni populations on projects for American or British clients. The firm did this at the local level, creating right-wing pages with vague names like Smith County Patriots or I Love My Country. Because of the way Facebook's recommendation algorithm worked, these pages would pop up in the feeds of people who had already liked similar content. When users joined CA's fake groups, it would post videos and articles that would further provoke and inflame them. Conversations would rage on the group page, with people commiserating about how terrible or unfair something was. CA broke down social barriers, cultivating relationships across groups. And all the while it was testing and refining messages, to achieve maximum engagement.

Now CA had users who (1) self-identified as part of an extreme group, (2) were a captive audience, and (3) could be manipulated with data. Lots of reporting on Cambridge Analytica gave the impression that everyone was targeted. In fact, not that many people were targeted at all. CA didn't need to create a big target universe, because most elections are zero-sum games: if you get one more vote than the other guy or girl, you win the election. Cambridge Analytica needed to infect only a narrow sliver of the population, and then it could watch the narrative spread.

Once a group reached a certain number of members, CA would set up a physical event. CA teams would choose small venues – a coffee shop or bar – to make the crowd feel larger. Let's say you have a thousand people in a group, which is modest in Facebook terms. Even if only a small fraction shows up, that's still a few dozen people. A group of forty makes for a huge crowd in the local coffee shop. People would show up and find a fellowship of anger and paranoia. This naturally led them to feel like they were part of a giant movement, and it allowed them to further feed off one another's paranoia and fears of conspiracy. Sometimes a Cambridge Analytica staffer would act as a 'confederate' – a tactic commonly used by militaries to stir up anxieties in target groups. But most of the time, these situations unfolded organically. The invitees were selected because of their traits, so Cambridge Analytica knew generally how they would react to one another. The meetings took place in counties all across the United States, starting with the early Republican primary states, and people would get more and more fired up at what they saw as 'us *vs* them.' What began as their digital fantasy, sitting alone in their bedrooms late at night clicking on links, was becoming their new reality. The narrative was right in front of them, talking to them, live in the flesh. Whether or not it *was* real no longer mattered; that it *felt* real was enough.

Cambridge Analytica ultimately became a digitised, scaled and automated version of a tactic the United States and its allies have used in other countries. When I first started at SCL, the firm had been working on counter-narcotics programs in a South American country. The strategy was, in part, to identify targets to disrupt narcotics organisations from within. The first thing the firm would do was find the lowest-hanging fruit, meaning people who its psychologists reasoned would be more likely to become more erratic or paranoid. Then the firm would work on suggesting ideas to them: 'The bosses are stealing from you' or 'They're going to let you take the fall.' The goal was to turn them against the organisation, and sometimes, if a person hears something enough times, they come to believe it.

Once those initial individuals were sufficiently exposed to these new narratives, it would be time to have them meet one another so that they could form a group which could then organise. They would share rumours, working one another into deeper paranoia. That was when you introduced the next tier: people whose initial resistance to rumours had started to weaken. And this is how you gradually destabilise an organisation from the inside. CA wanted to do the same to America, using social media as the spearhead. Once a county-based group begins self-organising, you introduce them to a similar group in the next county over. Then you do it again. In time, you've created a statewide movement of neurotic, conspiratorial citizens. The alt-right.

Internal tests also showed that the digital and social ad content being piloted by CA was effective at garnering online engagement. Those being targeted online with test advertisements had their social profiles matched to their voting records, so the firm knew their names and 'real world' identities. The firm then began to use numbers on the engagement rates of these ads to explore the potential impact on voter turnout. One internal memo highlighted the results from an experiment involving registered voters who had not voted in the two previous elections. CA estimated that if only 25 per cent of the infrequent voters who began clicking on this new CA content eventually turned out to vote, they could increase statewide turnout for the Republicans in several key states by around 1 per cent, which is often the margin of victory in tight races. Steve Bannon loved this. But he wanted CA to go further – and darker. He wanted to test the malleability of the American psyche. He urged us to include what were in effect racially biased questions in our research, to see just how far we could push people. The firm started testing questions about black people – whether they were capable of succeeding in America without the help of whites, for example, or whether they were genetically predetermined to fail. Bannon believed that the civil rights movement had limited 'free thinking' in America. He was determined to liberate people by revealing what he saw as the forbidden truths about race.

Bannon suspected that there were swathes of Americans who felt silenced by the threat of being labelled 'racist.' Cambridge Analytica's findings confirmed his suspicion: America is filled with racists who remain silent for fear of social shunning. But Bannon wasn't just focused on his emerging alt-right movement; he also had Democrats in mind.

While 'typical Democrats' talk a good game when it comes to supporting racial minorities, Bannon detected an underlying paternalism that betrayed their professed wokeness. The party, he felt, was full of 'limousine liberals' – a term coined in the New York mayoral race of 1969 and instantly seized on by populists to denigrate do-gooder Democrats. These were the white Democrats who supported school busing but sent their own kids to majority-white private schools, or who professed to care about the inner city but lived in gated communities. 'The Dems always treat blacks like children,' Bannon said on one call. 'They put them in projects ... give them welfare ... affirmative action ... send white kids to hand out food in Africa. But Dems are always afraid to ask the question: *Why do those people need so much babysitting?*'

What he meant was that white Democrats revealed their prejudices against minorities without realising it. He posited that although these Democrats think that they *like* African Americans, they do not *respect* African Americans, and that many Democratic policies stemmed from an implicit acknowledgment that *those people* cannot help themselves. Speechwriter Michael Gerson perfectly encapsulated this idea with a phrase he coined for then-presidential candidate George W. Bush in 1999: 'The soft bigotry of low expectations.' According to this argument, Democrats were hand-holders, enablers of bad behaviour and poor testing results because they didn't actually believe that minority students could do as well as their non-minority peers.

Bannon had a starker, more aggressive take on this idea: he believed the Democrats were simply using American minorities for their own political ends. He was convinced that the social compact that emerged after the civil rights movement, where Democrats benefitted from African American votes in exchange for government aid, was not born out of any moral enlightenment, but instead out of shrewd calculation. In his framing, the only way the Democrats could defend what he saw as the inconvenient truths of this social compact was through *political correctness*. Democrats subjected 'rationalists' to social shame when they spoke out about this 'race reality.'

'Race realism' is the most recent spin on age-old tropes and theories that certain ethnic groups are genetically superior to others. Race realists believe, for example, that black Americans score lower on standardised tests not because the tests are skewed, or because of the long history of oppression and prejudice that blacks must overcome, but because they're inherently less intelligent than white Americans. It's a pseudoscientific notion, embraced by white supremacists, with roots in the centuries-old 'scientific racism' that underlies, among other disasters of human history, slavery, apartheid and the Holocaust. The alt-right, led by Bannon and Breitbart, adopted race realism as a cornerstone philosophy.

If Bannon were to succeed in his quest for liberation of his 'free thinkers,' he needed a way of inoculating people from political correctness. Cambridge Analytica began studying not only *overt racism* but racism in its many other incarnations. When we think about racism, we often think of overt hatred. But racism can persist in different ways. Racism can be *aversive,* where a person consciously or subconsciously avoids a racial group (e.g., gated communities, sexual and romantic avoidance, etc.), and racism can be *symbolic,* where a person holds negative evaluations of a racial group (e.g., stereotypes, double standards, etc.). However, because the label 'racism' can hold such social stigma in modern America, we found that white people often ignore or discount their internalised prejudices and react strongly to any inference that they hold such beliefs.

This is what is known as 'white fragility': white people in North American society enjoy environments insulated from racial disadvantages, which fosters an expectation among white people of racial comfort while lowering their ability to tolerate racial stress. In our research, we saw that white fragility prevented people from confronting their latent prejudices. This cognitive dissonance also meant that subjects would often amplify their responses expressing positive statements toward minorities in an effort to satiate their self-concept of 'not being racist'. For example, when presented with a series of hypothetical biographies with photos, some respondents who scored higher in prior implicit racial bias testing would rate minority biographies higher than identical white biographies. *See? I scored the black person higher, because I am not racist.*

This cognitive dissonance created an opening: many respondents were reacting to their own racism not out of concern about how they may be contributing to structural oppression, but rather to protect their own social status. For Bannon, this was enough to convince him that his theory about Democrats was true – that they just pay lip service to minorities, but deep down they are just as racist as anyone else in America. The difference was who was living in what 'reality.'

BANNON ENVISIONED A VEHICLE to help white racists move past all this and become liberated 'free thinkers'. In 2005, when Bannon started at IGE, the Hong Kong-based gaming company, the firm employed a factory of low-wage Chinese gamers to play *World of Warcraft* in order to win items in the game. Instead of trading them, or selling them through the game's interface, which was allowed, IGE would sell the digital assets to western players for a profit. This activity was largely seen by other players as cheating, and a civil suit and backlash online against the firm ensued. It's possible this was Bannon's early exposure to the rage of online communities; some of the commentary was reportedly 'anti-Chinese vitriol'. Bannon became a regular reader of Reddit and 4chan and began to see the hidden anger that comes out when people are anonymous online. To him, they were revealing their true selves, unfiltered by a 'political correctness' that was preventing them from speaking these 'truths' in public. It was through the process of reading these forums that Bannon realised he could harness them and their anonymous swarms of resentment and harassment.

This was especially true after Gamergate, in the late summer of 2014, right before Bannon was introduced to SCL. In many ways, Gamergate created a conceptual framework for Bannon's alt-right movement, as he knew there was an undercurrent populated by millions of intense and angry young men. Trolling and cyberbullying became key tools of the alt-right. But Bannon went deeper and had Cambridge Analytica scale and deploy many of the same tactics that domestic abusers and bullies use to erode stress resilience in their victims. Bannon transformed CA into a tool for automated bullying and scaled psychological abuse. The firm started this journey by identifying a series of cognitive biases that it hypothesised would interact with latent racial bias. Over the course of many experiments, we concocted an arsenal of psychological tools that could be deployed systematically via social media, blogs, groups and forums.

Bannon's first request of our team was to study who felt oppressed by political correctness. Cambridge Analytica found that, because people often overestimate how much others notice them, spotlighting socially uncomfortable situations was an effective prime for eliciting bias in target cohorts, such as when you get in trouble for mispronouncing a foreign-sounding name. One of the most effective messages the firm tested was getting subjects to *'imagine an America where you can't pronounce anyone's name.'* Subjects would be shown a series of uncommon names and then asked, *'How hard is it to pronounce this name? Can you recall a time where people were laughing at someone who messed up an ethnic name? Do some people use political correctness to make others feel dumb or to get ahead?'*

People reacted strongly to the notion that 'liberals' were seeking new ways to mock and shame them, along with the idea that political correctness was a method of persecution. An effective Cambridge Analytica technique was to show subjects blogs that made fun of white people like them, such as *People of Walmart*. Bannon had been observing online communities on places like 4chan and Reddit for years, and he knew how often subgroups of angry young white men would share content of 'liberal elites' mocking 'regular' Americans. There had always been publications that parodied the 'hicks' of flyover country, but social media represented an extraordinary opportunity to rub 'regular' Americans' noses in the snobbery of coastal elites.

Cambridge Analytica began to use this content to touch on an implied belief about racial competition for attention and resources – that race relations were a zero-sum game. The more they take, the less you have, and they use political correctness so you cannot speak out. This framing of political correctness as an identity threat catalysed a 'boomerang' effect in people where counternarratives would actually strengthen, not weaken, the prior bias or belief. This means that when targets would see clips containing criticism of racist statements by candidates or celebrities, this exposure would have the effect of further entrenching the target's racialised views, rather than causing them to question those beliefs. In this way, if you could frame racialised views through the lens of identity prior to exposure to a counternarrative, that counternarrative would be interpreted as an attack on identity instead. What was so useful for Bannon was that it in effect inoculated target groups from counternarratives criticising ethno-nationalism. It created a wicked reinforcement cycle in which the cohort would strengthen their racialised views when they were exposed to criticism. This may be in part because the area of the brain that is most highly activated when we process strongly held beliefs is the same area that is involved when we think about who we are and our identity. Later, when Donald Trump was aggressively criticised in the media for racist or misogynist statements, these critiques likely created a similar effect, where the criticism of Trump strengthened the resolve of supporters who would internalise the critique as a threat to their very identity.

By making people angry in this way, CA was following a fairly wide corpus of research showing that anger interferes with information seeking. This is why people can 'jump to conclusions' in a fit of rage, even if later they regret their decisions. In one experiment, CA would show people on online panels pictures of simple bar graphs about uncontroversial things (e.g., the usage rates of mobile phones or sales of a car type) and the majority would be able to read the graph correctly. However, unbeknownst to the respondents, the data behind these graphs had actually been derived from politically controversial topics, such as income inequality, climate change or deaths from gun violence. When the labels of the *same graphs* were later switched to their actual controversial topic, respondents who were made angry by identity threats were more likely to misread the relabelled graphs that they had previously understood.

What CA observed was that when respondents were angry, their need for complete and rational explanations was also significantly reduced. In particular, anger put people in a frame of mind in which they were more indiscriminately punitive, particularly to out-groups. They would also underestimate the risk of negative outcomes. This led CA to discover that even if a hypothetical trade war with China or Mexico meant the loss of American jobs and profits, people primed with anger would tolerate that domestic economic damage if it meant

in the South all but unable to vote. And the Ku Klux Klan, which had virtually disappeared just after the Civil War, enjoyed a resurgence in the early twentieth century, in part by presenting itself as a national patriotic organisation.

The Civil Rights Act of 1964 and the Voting Rights Act of 1965 represented a huge leap forward for the rights of American blacks. These sweeping sets of laws promised to right many of the wrongs that had been perpetrated against the black community for so many years by ensuring voting rights, mandating desegregation of public facilities, and instituting equal employment opportunity and non-discrimination in federal programs. They also opened a new chapter in the politics of shamelessly stoking white fear.

In the late 1960s, Richard Nixon's 'southern strategy' fuelled racial fear and tensions in order to shift white voters' allegiance from the Democrats to the Republicans. Nixon ran his 1968 presidential campaign on the twin pillars of 'states' rights' and 'law and order' – both of which were obvious, racially coded dog whistles. In his 1980 campaign, Ronald Reagan repeatedly invoked the 'welfare queen' – a black woman who supposedly was able to buy a Cadillac on government assistance. In 1988, George H. W. Bush's campaign ran the infamous Willie Horton ad, terrifying white voters with visions of wild-haired black criminals running amok.

Steve Bannon aimed to affirm the ugliest biases in the American psyche and convince those who possessed them that they were the victims, that they had been forced to suppress their true feelings for too long. Deep within America's soul lurked an explosive tension. Bannon had long sensed this, and now he had the data to prove it. History, Bannon was convinced, would prove to be on his side, and the right tools would hasten his prophecy. Young people, with their lack of opportunities stemming from a corpulent state and a corrupt finance system, were primed to rebel. They just did not know it yet. Bannon wanted them to understand their role in his prophecy of revolution – that they would lead a generational 'turning' of history and become the 'artists' who would redraw a new society filled with meaning and purpose after its 'great unravelling.' Major figures in history, he said, were artists: Franco and Hitler were painters, while Stalin, Mao and bin Laden were all poets. He understood that movements adopt a new aesthetic for society. Bannon asked why dictators always lock up the poets and artists first. Because they are often artists themselves. And for Bannon, this movement was primed to become his great performance. It was the fulfilling of his prophecy by making real the narratives of his favourite books, like *The Fourth Turning*, which predicts an impending crisis followed by a forgotten generation rising up in rebellion, or *The Camp of the Saints*, where western civilisation collapses from the weight of caravans of immigrant invaders.

But Bannon needed an army to unleash chaos. For him, this was an insurgency, and to inspire total loyalty and total engagement, he was prepared to use any narrative that worked. The exploitation of cognitive biases, for Bannon, was simply a means of 'de-programming' his targets from the 'conditioning' they had endured growing up in a vapid and meaningless society. Bannon wanted his targets to 'discover themselves' and 'become who they really were'. But the tools created at Cambridge Analytica in 2014 were not about self-actualisation; they were used to accentuate people's innermost demons in order to build what Bannon called his 'movement'. By targeting people with specific psychological vulnerabilities, the firm victimised them into joining what was nothing more than a *cult* led by false prophets, where reason and facts would have little effect on its new followers, digitally isolated as they now were from inconvenient narratives.

In the last discussion I ever had with Bannon, he told me that to fundamentally change society, 'you have to break everything'. And that's what he wanted to do – to fracture 'the establishment'. Bannon faulted 'big government' and 'big capitalism' for suppressing the randomness that is essential to human experience. He wanted to liberate the people from a controlling administrative state that made choices for them and thus removed purpose from their lives. He wanted to bring about chaos to end the tyranny of certainty within the administrative state. Steve Bannon did not want, and would not tolerate, the state dictating America's destiny.

8

FROM RUSSIA WITH LIKES

KEEPING TRUE TO ITS ORIGINS IN FOREIGN INFORMATION operations, there were new characters arriving at Cambridge Analytica's London office almost daily. The firm became a revolving door of foreign politicians, fixers, security agencies and businessmen with their scantily clad private secretaries in tow. It was obvious that many of these men were associates of Russian oligarchs who wanted to influence a foreign government, but their interest in foreign politics was rarely ideological. Rather, they were usually either seeking help to stash money somewhere discreet, or to retrieve money that was sitting in a frozen account somewhere in the world. Staff were told to just ignore the comings and goings of these men and not ask too many questions, but staff would joke about it on internal chat logs, and the visiting Russians in particular were usually the more eccentric variety of clients we would encounter. When the firm would conduct internal research on these potential clients, we would hear through the grapevine about the amusing hobbies or bizarre sexual escapades these powerful men would get up to. And I did, admittedly, turn a blind eye to the firm's meetings with suspicious-looking clients. I knew it would just get me in trouble with Nix if I asked too many questions that he didn't care for. But at the time, in spring 2014, just two years before any Russian disinformation efforts hit the US presidential election, there wasn't anything innately suspicious about these Russians beyond the typical bread-and-butter shadiness that the firm engaged in. That is, except for one prospective client that CA executives became both very giddy and unusually elusive about.

In the spring of 2014, the large Russian oil company Lukoil contacted Cambridge Analytica and began asking questions. At first, Nix handled the conversations, but soon the oil executives wanted answers that he was incapable of providing. He sent Lukoil CEO Vagit Alekperov a white paper I'd written about Cambridge Analytica's data targeting projects in the United States, after which Lukoil asked for a meeting. Nix said that I should come along. 'They understand behavioural micro-targeting in the context of elections (as per your excellent document/white paper) but they are failing to make the connection between voters and their consumers,' he wrote in an email.

Well, I was failing to make that connection, too. Lukoil was a major force in the global economy – the largest privately owned company in Putin's kleptocracy – but I couldn't see an obvious link between a Russian oil company and CA's work in the United States. And Nix was of no help. 'Oh, you know how these things are,' he told me. 'You just lift your skirt a little, and then they give you money.' In other words, he wasn't interested in the details. If Lukoil wanted to pay for our data, why should we care what they did with it?

Shortly after the first Lukoil approach, a 2014 memo on CA's internal capacity was drafted and sent to Nix. The briefing discussed in euphemistic terms what the firm, at least in theory, was capable of setting up were there to be a project that needed special intelligence services or scaled disinformation operations on social media. (As the memo was internal, it referenced SCL; Cambridge Analytica was merely a front-facing brand for American clients that was entirely staffed by SCL personnel.) 'SCL retains a number of retired intelligence and security agency officers from Israel, USA, UK, Spain & Russia each with extensive technical and analytical experience,' the memo read. 'Our experience shows that in many cases utilising social media or "foreign" publications to "expose" an opponent is often more effectual than using potentially biased local media channels.' The memo discussed 'infiltrating' opposition campaigns using 'intelligence nets' to obtain 'damaging information' and creating scaled networks of 'Facebook and Twitter accounts to build credibility and cultivate followers.' For many of SCL's clients, this was a standard offer – private espionage, stings, bribes, extortion, infiltrations, honey traps and disinformation spread through fake accounts on social media. For the right price, SCL was willing to do whatever it took to win an election. And now, armed with even more extensive data sets and AI capabilities, and millions invested, the newly formed Cambridge Analytica was looking to take this further.

The Lukoil execs came to London, where Nix had prepared a pitch deck of slides for the meeting. I sat back in my chair, curious to discover what the hell he was actually pitching. The first couple of slides outlined an SCL project in Nigeria aimed at undermining voters' confidence in civic institutions. Labelled 'Election: Inoculation,' the material described how to spread rumours and disinformation to sway election results. Nix played videos of emotional voters convinced that the upcoming Nigerian election would be rigged.

'We made them think that,' he said with delight.

The next set of slides described how SCL had worked to fix elections in Nigeria, complete with videos of voters saying how worried they were about rumours of violence and upheaval. 'And we made them think that too,' Nix said.

I watched in silence as these Russian executives took notes, casually nodding along as if what they were seeing was totally routine. Next, Nix showed them slides about our data assets. But we didn't have data assets in Russia or the Commonwealth of Independent States (CIS) markets that Lukoil primarily operated in, and our largest data set covered America. Then he started talking about microtargeting and AI, and what Cambridge Analytica was doing with the data in our possession.

I was still at a loss. At the end of the presentation, the executives asked me what I thought, and I fumfered a bit, saying, 'Well, we have a diverse set of experiences and data in many places . . . So why exactly are you interested in all this?'

One of them responded that they were still figuring that out, and we should continue telling them more about what data and capacity CA had. But I was the one who needed answers. Why would a Russian oil company with virtually no presence in the United States want access to our US data assets? And if this was a commercial project, why was Alexander showing them slides about disinformation in Africa?

But it wasn't simply the internal data assets that were shown to the firm's clients. The firm was eager to show off to prospective clients how much it knew about internal US military operations. In another meeting, an internal slide deck created by the US Air Force Targeting Center in Langley, Virginia, which the firm somehow had accessed, outlined for some prospective clients how the United States was already 'incorporating socio cultural behaviour factors into operational planning' in order to gain the 'ability to "weaponeer" targets' and amplify non-kinetic force against American adversaries. Nix remained coy about his plans. This struck me as against type – how many times had I seen him banter about the firm bribing ministers or setting up honey traps? But he couldn't – or wouldn't – explain why we kept communicating with this 'client'. And during the discussions, he kept telling them, 'We've already got guys on the ground.'

A FEW MONTHS BEFORE the first round of Lukoil meetings, Cambridge Analytica had connected with a man named Sam Patten, who had lived a colourful life as a political operative for hire all over the world. In the 1990s, Patten worked in the oil sector in Kazakhstan before moving into Eastern European politics. When CA hired him, he had just finished a project for pro-Russian political parties in Ukraine. At the time, he was working with a man named Konstantin Kilimnik, a former officer of Russia's Main Intelligence Directorate (the GRU). Although Patten denies that he gave his Russian partner any data, it was later revealed that Paul Manafort, who was for several months Donald Trump's campaign manager, did pass along voter polling data to Kilimnik in a separate instance. Patten and Kilimnik had met in Moscow in the early 2000s and later worked in Ukraine for Paul Manafort's consultancy. The two became formal business partners soon after Patten was brought onto CA.

Patten was a perfect fit to navigate the world of shady international influence operations. He was also well connected among the growing number of Republicans joining Cambridge Analytica, so he was initially assigned to work in the United States. Patten was tasked with managing the logistics of research operations in America, including focus groups and data collection, and writing some of the polling questions. In spring 2014, he started working in Oregon, taking over for some of Gettleson's projects conducting social and attitudinal research on American citizens.

Soon enough, weird questions began popping up in our research. One day I was in my London office, checking reports from the field, when I noticed a project involving Russia-oriented message testing in America. The US operation was growing rapidly, and several new people had been brought in to manage the surge in assignments, so it was hard to keep track of every research stream. I thought that maybe someone had started exploring Americans' views on international topics. But when I searched our repository of questions and data, I could only find data being collected on Russia. Our team in Oregon had started asking people, 'Is Russia entitled to Crimea?' and 'What do you think about Vladimir Putin as a leader?' Focus group leaders were circulating various photos of Putin and asking people to indicate where he looked strongest. I started watching video recordings of some of the focus groups – and they were strange. Photos of Vladimir Putin and Russian narratives were projected on the wall, and the interviewer was asking groups of American voters how it made them feel to see a strong leader.

What was interesting was that even though Russia had been a US adversary for decades, Putin was admired for his strength as a leader.

'He has a right to protect his country and do what he thinks is best for his country,' said one participant as others nodded in agreement. Another told us that Crimea was Russia's Mexico, but that, unlike Obama, Putin was taking action. As I sat alone in the now dark office, watching bizarre clips of Americans discussing Putin's claim to Crimea, I wanted answers. Gettleson was in America at the time. When he answered the phone, I asked if he could enlighten me about who had authorised a research stream on Putin. He had no idea. 'It just showed up,' he said, 'so I assumed it was approved by someone.'

Patten's interest in Eastern European politics crossed my mind, but I didn't give it a lot of thought. In August 2014, a Palantir staff member sent an email to the data science team with a link to an article about Russians stealing millions of internet browsing records. 'Talk about acquiring data!' they joked. Two minutes later, one of our engineers responded, 'We can exploit similar methods.' Maybe he was joking, maybe he wasn't, but the firm had already contracted former Russian intelligence officers for other projects, as the memo to Nix highlighted.

Kogan, the project's lead psychologist since May 2014, was making trips to St Petersburg and Moscow. He was not forthcoming about his projects in Russia, but I knew he was working on psychometric profiling of social media users. The research Kogan was doing in Russia was focused on identifying disordered people and exploring their potential for trolling behaviour on social networks. His research at St Petersburg State University, funded by a Russian government research grant, examined connections between dark-triad personality traits and engagement in cyberbullying, trolling and cyber stalking. The research also explored political themes on Facebook, finding that high scorers in psychopathy were most likely to post about authoritarian political issues. In conjunction with clinical and computational psychologists, Kogan worked with the 'data of Facebook users from Russia and the USA by means of a special web-application,' according to one of the research briefings from his Russian research team. By late summer, Kogan was delivering lectures in Russia on the potential political applications of social media profiling. I remember him mentioning to me that there was 'overlap' between his work in St Petersburg and at Cambridge Analytica, but this could have been a coincidence. My own personal belief, which I expressed to Congress, was that Kogan was not ill-intentioned, but merely careless and naïve. Objectively, the security for the data was poor.

Even before Kogan came along, CA's parent company, SCL, had deep experience disseminating propaganda online, but Kogan's research was well suited to targeting voters with authoritarian personality traits, identifying narratives that would activate their support. After Kogan joined Cambridge Analytica's project, CA's internal psychology team started replicating some of his research from Russia: profiling people who were high in neuroticism and dark-triad traits. These targets were more impulsive and more susceptible to conspiratorial thinking, and, with the right kind of nudges, they could be lured into extreme thoughts or behaviour.

—

IT TAKES TOXIC LEADERS to create a toxic enterprise, and I think Cambridge Analytica reflected the character of Nix. Along with the obvious delight he took in intimidation, Nix possessed an uncanny gift for finding just the spot where his malice would do the most damage. He wouldn't stop calling me a 'gimp' or 'spaz case', for example, because he knew it made me feel weak. But he knew it would only make me work harder for him. As much as I resented him, for some reason I became determined to prove him wrong. The constant abuse came with an explanation: only 'the truth' could motivate someone to rise to Nix's standards. He also made sport of belittling staff, blowing through the office like some irritable tornado, tossing out insults as he passed.

On one occasion we thought Nix was going to hurt someone. I can't even remember what provoked him this time, but for whatever reason, he flipped out and pushed everything off one of the interns' desks. Nix was screaming, leaning in so closely that tiny droplets of spit were hitting the intern's cheek. Tadas Jucikas, who was the largest of us, got up and walked over. 'Alexander, it sounds like you need a drink,' he said. 'How about you join me at the club.' After Nix left, the intern just sat there, breathing heavily, until another colleague suggested he leave for the day. We all cleaned up the mess Nix had made before he returned in a much more jovial mood, as if nothing had happened.

Sometimes he would blame the victim after losing his temper. 'You always make me yell,' he'd say, as if he was not in control of his own voice. What disturbed me most was when he denied a tantrum even as I was still reeling from the effects. There is something quite powerful in being told, flatly, that the thing that upset you never happened; eventually you start to worry that you've gone mad. 'You need to grow up and be less sensitive,' Nix would say. 'I can't trust you if you keep telling me that I lost my temper.'

We had one huge blowout that ended up having both short-term and long-term consequences. When Cambridge Analytica was officially formed, I kept refusing to sign my contract. Signing could have granted me shares, but I was nervous about making a long commitment to the firm. A voice in the back of my head warned against it.

The delay made Nix furious. Finally he snapped, locking me in a room, where he screamed and berated me. When that didn't yield the desired result, he flipped over the chair beside me. As soon as he opened the door, I rushed out of the office and didn't return for two weeks. We both knew that he needed me more than I needed him, because I was the only one who could build what he had promised the Mercers. But he was still too stubborn and haughty to tell me he was sorry, so after a while he asked Jucikas to convey an apology to me. I reluctantly came back to work, but I still refused to sign the contract.

CA's client list eventually grew into a who's who of the American right wing. The Trump and Cruz campaigns paid more than $5 million apiece to the firm. The US Senate campaigns of Roy Blunt of Missouri and Tom Cotton of Arkansas became clients. And, of course, there was the losing House bid of Art Robinson, the Oregon Republican who collected piss and church organs. In the autumn of 2014, Jeb Bush paid a visit to the office. Despite having received millions from Mercer, Nix never bothered to learn much about US politics, so he asked Gettleson to join him. Bush, who had come alone, began by telling Nix that if he decided to run for president, he wanted to be able to do it *on his terms,* without having to 'court the crazies' in his party.

'Of course, of course,' Nix answered, signalling his intention to bluff and bullshit his way through the entire meeting. When it was over, he was so excited at the possibility of signing up another big American client, he insisted on immediately calling the Mercers with the good news, having apparently forgotten that the Mercers had told him on countless occasions of their support for Ted Cruz. Nix put Rebekah Mercer on speakerphone so that everyone could hear her reaction to the amazing meeting he'd just had.

'We've just had Governor Jeb Bush in the office, and he wants to work with us. What do you think of that?' he said proudly. After a pause, Rebekah replied flatly, 'Well, I hope you told him very clearly that that's never happening.' Then she hung up. Brutal.

And it wasn't just presidential hopefuls who sought CA's help. For evangelical leader Ralph Reed, Nix planned a lunch at the grand dining hall of the Oxford and Cambridge Club, on Pall Mall. Reed spent two hours describing his objectives and outlining how CA could help re-instil morality in an America fighting over same-sex marriage and other cultural issues. Nix left the meeting a little drunk. Back at the office, he announced in his typical outrageous fashion to everyone, 'Well, there's a closet case if I've ever seen one.'

For most of the time I was at SCL and Cambridge Analytica, none of what we were doing felt real, partly because so many of the people I met seemed almost cartoonish. The job became more of an intellectual adventure, like playing a video game with escalating levels of difficulty. *What happens if I do this? Can I make this character turn from blue to red, or red to blue?* Sitting in an office, staring at a screen, it was easy to spiral down into a deeper, darker place, to lose sight of what I was actually involved in.

But eventually, I couldn't ignore what was right in front of my eyes. Weird PACs (Political Action Committee) started showing up. The super PAC of future national security adviser John Bolton paid Cambridge Analytica more than $1 million to explore how to increase militarism in American youth. Bolton was worried that millennials were a 'morally weak' generation that would not want to go to war with Iran or other 'evil' countries.

Nix wanted us to start using disguised names for any client research in the United States and to state that the research was being conducted for the University of Cambridge. I tried to put a stop to this in an email to staff: 'You cannot lie to people,' I wrote, citing possible legal consequences. The warning was ignored.

At this point I was feeling more and more as if I was a part of something that I did not understand and could not control, and that was, at its core, deeply unsavory. But I also felt lost and trapped. I started going out on all-night binges at late-night clubs or raves. A couple of times, I left the office in the evening, went out all night, and ended up coming back in without actually going to bed. My friends in London noticed that I was no longer myself. Gettleson finally said to me, 'You don't look well, Chris. Are you okay?' I wasn't; I was despondent. There were days when I wanted to scream back at Nix, but something stopped me. I would go out, sometimes alone, and the loud music and constant contact with other dancing bodies made me feel as if I was still here and this wasn't some dream. And if the music is loud enough, you can scream at the world and no one notices.

OUR WORK AT CAMBRIDGE ANALYTICA seemed to grow more nefarious every day. One project was described in CA correspondence as a 'voter disengagement' (i.e., voter suppression) initiative targeting African Americans. Republican clients were worried about the growing minority vote, especially in relation to their aging white base, and were looking for ways to confuse, demotivate and disempower people of colour. When I found out that CA was beginning a voter suppression project, it really hit home. I thought about all the times I had gone to rallies in 2008, when Barack Obama was running, and I started to ask myself, *How in the hell did I end up here?* I told one of the new managers that, regardless of what the client wanted, it could be illegal to work on a project with the objective of voter suppression. Once again, I was ignored. I called the firm's US lawyers in New York and left a message asking them to call me back, but they never did.

IN JULY 2014, I was copied on a confidential memo sent to Bannon, Rebekah Mercer and Nix by Bracewell & Giuliani, the law firm of Rudy Giuliani. Cambridge Analytica had sought advice on US law regarding foreign influence on campaigns. The memo outlined the Foreign Agents Registration Act and was emphatically clear: foreign nationals are strictly prohibited from managing or influencing an American campaign or PAC at the local, state or federal level. The memo recommended that Nix immediately recuse himself from substantial management of Cambridge Analytica until 'loopholes' could be explored. The Bracewell & Giuliani memo suggested 'filtering' the work of CA's foreign nationals through US citizens. After reading the memo, I pulled Nix into a meeting room to urge him to heed the warning.

Instead, Cambridge Analytica started to require foreign staff members to sign a waiver before flying to America, accepting liability for any breach of election law. They were not informed of the advice from Giuliani's firm. It set me off. I unloaded on Nix.

'What if they get prosecuted, Alexander?' I shouted. 'That will be on you.'

'It's *their* responsibility, not mine, to know what the rules are,' he replied. 'They are adults. They can make decisions for themselves.'

But it was his decisions I was worried about, and I wasn't alone. Filling me in on some new projects, a colleague on the psychology team shared similar concerns about how this research could be used to amplify, rather than moderate, racism in the populations Cambridge Analytica was focusing on. 'I don't think we should keep doing this research,' he said.

Originally, race was one of many topics the firm began exploring. This in itself was not unusual, as racial conflicts have played a significant role in American culture and history. Psychologists on the projects initially assumed this research would be used either for passive information about the biases of the population or even to help reduce their effects. But, lacking the kind of traditional ethics review that is a prerequisite of academic research, there was never consideration of how this research could be misused – no one thought about how this could go wrong.

I knew Bannon would go on rants about how America was changing too much, his prophetic notion of an impending great conflict, or his misreading of dharma in Hinduism, which bordered on fetishistic Orientalism. But many of us on the CA research teams brushed him off as just another eccentric person we had to placate in the bizarre world we worked in. Many of the CA staff had experience working in far more extreme circumstances on old SCL information operations projects around the world, so by comparison Bannon felt quite tame.

But as CA rapidly grew after Mercer's investment, I hadn't fully grasped the scale of the race projects we were involved in. The new managers that Nix and Bannon hired started excluding me from meetings, and I stopped getting automatically invited to project planning meetings. I thought this was another power trip by Nix, so I simply felt annoyed rather than suspicious. But one of the psychologists on the team started coming to me to show me some of the new race projects. He showed me the master document of research questions that were being fielded in America, and my stomach dropped when I started reading. We were testing how to use cognitive biases as a gateway to move people's perceptions of racial out-groups. We were using questions and images clearly designed to elicit racism in our subjects. As I watched a video of a man who was a participant in one of the field experiments, who'd been provoked by a CA researcher's guided questioning into spasms of rage, racist insults flying from his mouth, I started to confront what I was helping to build.

In our invasion of America, we were purposefully activating the worst in people, from paranoia to racism. I immediately wondered if this was what Stanley Milgram felt like watching his research subjects. We were doing it in service to men whose values were in total opposition to mine. Bannon and Mercer were more than happy to hire the very people they sought to oppress – queers, immigrants, women, Jews, Muslims and people of colour – so that they could weaponise our insights and experiences to advance these causes. I was no longer working at a firm that fought against radical extremists who shackled women, brutalised nonbelievers and tortured gays; I was now working *for* extremists who wanted to build their very own dystopia in America and Europe. Nix knew this and didn't even care. For the cheap thrill of sealing another deal, he had begun entertaining bigots and homophobes, expecting his staff not only to look the other way, but for us to betray *our own people*.

In the end, we were creating a machine to contaminate America with hate and cultish paranoia, and I could no longer ignore the immorality and illegality of it all. I did not want to be a collaborator.

Then, in August 2014, something terrible happened. A veteran SCL staffer, a longtime friend and confidant of Nix's, returned from Africa severely ill with malaria. He came into the office red-eyed and sweating profusely, slurring his words and talking nonsense. After Nix shouted at him for being late, the rest of us urged him to go to the hospital. But before he could be seen at the hospital, he collapsed and tumbled down a flight of stairs, smashing his head hard on the concrete. He slipped into a coma. His brain swelled and part of his skull was removed. His doctors worried that his cognitive functioning might never be the same.

After Nix returned from visiting the hospital, he asked HR for guidance on liability insurance and how long he had to keep paying his loyal friend, still in a coma and missing part of his skull. This seemed callous in the extreme. It was in that moment that I realised Nix was a monster. Worse, I knew he wasn't alone.

Bannon was also a monster. And soon enough, were I to stay, I worried that I would become a monster, too.

The social and cultural research I'd been enjoying only a few months before had given birth to this *thing* – and it was terrifying. It is hard to explain what the atmosphere was like, but it was as if everyone had become detached from the realities of what we were doing. But I had snapped out of the daze and was now watching a revolting idea become real. My head cleared, and the real-world consequences of Nix's evil dreams began to haunt me. Late into the evening, unable to sleep, I would stare at the ceiling, my thoughts stalled between agony and bewilderment. One night, I called my parents in Canada, at 3 a.m. their time, to ask for advice. 'Read the signs,' they said. 'If you can't sleep – if you're making calls at all hours in a panic for answers – then you know what you should do.'

I told Nix that I was leaving. I wanted to get away from his psychopathic vision – and that of Bannon – as fast as I could. Otherwise I risked catching the same disease of mind and spirit.

Nix countered by appealing to my sense of loyalty. He made me think that I would be a bad person if I abandoned my friends at the firm. I was the one who had recruited people to work on Bannon's project. They trusted me, and I didn't want to betray them.

'Chris, you *cannot* leave me alone here with Nix,' said Mark Gettleson, who had joined the company in large part to work with me. 'If you go, I go.'

I didn't like the idea of walking out on my friends and colleagues, but I hated what Cambridge Analytica had become and what it was doing in the world. I told Nix that we could discuss how I would be phased out, but that I was definitely leaving. He did what came naturally – he took me to lunch.

The restaurant was in Green Park, not far from Buckingham Palace.

As soon as we sat down, Nix said, 'All right, then. I was expecting we would have this conversation eventually. How much do you want?'

I told him it wasn't about the money.

'Come on,' he said. 'I've run this firm long enough to know it's always about the money.'

He mentioned that I'd never asked for a raise, unlike some of my colleagues, despite how little he'd been paying me. And it was true: I had one of the lower salaries in the office, about half of what others were making, whereas recruits for Project Ripon were taking home triple to quadruple that. When I shook my head, Nix said, 'Fine. I'll just double your salary. That should do it.'

'Alexander,' I said, 'this is not some game I'm playing. I am leaving. I don't want to work here anymore. I'm done with whatever this is.' My tone deepened, and he seemed to finally realise I meant it, because then he leaned toward me and said, 'But Chris, this is your *baby*. And I *know* you. You wouldn't abandon *your* baby out in the streets, would you?' He must have sensed an opening, because he took the idea and ran with it. 'It's just been *born*. Don't you want to see it grow up? To know what school it goes to? If we can get it into Eton? To see what it accomplishes in life?'

He seemed pleased with the metaphorical flourish, but I wasn't the least bit moved. I told him that I felt less like a father than a sperm donor, with no power to keep the baby from growing into a hateful child. Nix quickly pivoted, suggesting we set up a Cambridge Analytica 'fashion division'.

'Jesus Christ, Alexander. Are you serious? Psychological warfare, the Tea Party ... *and fucking fashion trends*? No, Alexander. That's ridiculous.'

Finally, he got angry. 'You'll end up being the fifth Beatle,' he said.

The fifth beetle? I thought. Was this some kind of Egyptian parable? Something to do with scarabs? What in the world was he talking about? Not until later did I realise he was talking about the band that had formed three decades before I was born.

Even after I met him halfway, agreeing to stay on until the midterms in early November, Nix continued to insist that I was making a mistake.

'You don't even understand the enormity of what you have created here, Chris,' he said. 'You're only going to understand it when we're all sitting in the White House – every single one of us, except for you.'

Seriously? Even for Nix, this was grandiose. I could have had a nameplate in the West Wing, he told me. I was too stupid to realise what I was giving up.

'If you leave, that's it,' he said. 'Do not come back.'

I stayed for less than a year after Bannon took over and unleashed chaos. But looking back, I struggle to understand how I could have stayed even that long. Every day, I overlooked, ignored or explained away warning signs. With so much intellectual freedom, and with scholars from the world's leading universities telling me we were on the cusp of 'revolutionising' social science, I had gotten greedy, ignoring the dark side of what we were doing. Many of my friends did the same. I tried to convince Kogan to leave, too, and even when he conceded that the project could become an ethical quagmire, he decided to continue collaborating with Cambridge Analytica after I left. When I found out Kogan was staying, I refused to help him acquire more data sets for his projects, as I was worried that any new data I got for him could end up in the hands of Nix, Bannon and Mercer. What in my mind was meant to become an academic institute was becoming just another player in Cambridge Analytica's expanding web of partners. When I refused to continue helping Kogan, he demanded that I get rid of any data I had received from him, which I did. But this came at a huge cost for me personally, as Kogan had specifically added in fashion and music questions to the panels so I could incorporate the survey responses into my PhD thesis on trend forecasting. With the basis of my academic work now gone, I knew I would have to give up my PhD, which had become the only thing that was keeping me going. But what bothers me most was how I had let Nix dominate me. I let him pick away at every insecurity and vulnerability I had, and then, in service to him, I picked away at the insecurities and vulnerabilities of a nation. My actions were inexcusable, and I will always live with the shame.

JUST BEFORE I LEFT Cambridge Analytica, the firm was planning more election work in Nigeria. As Nix had explained to Lukoil in his presentation about rumour campaigns, the African nation was familiar territory. Cambridge Analytica knew that numerous foreign interests had a hand in African elections, making it unlikely that anyone would care what the firm was up to – it's Africa, after all. Following the frenzy of decolonisation in the 1960s, many western powers still felt entitled to interfere with their former African territories; the only difference now was the need for a measure of discretion. Europe had been built on African oil, rubber, minerals and labour, and the mere fact of a former colony's political independence was not going to change that.

With the Nigeria project, Cambridge Analytica pushed itself even deeper into psychologically abusive experiments. At the same hotel where Cambridge Analytica set up camp, Israeli, Russian, British and French 'civic engagement' projects operated behind fig-leaf cover stories. The unspoken belief shared by all: foreign interference in elections does not matter if those elections are African.

The company was working nominally in support of Goodluck Jonathan, who was running for reelection as the president of Nigeria. Jonathan, a Christian, was running against Muhammadu Buhari, who was a moderate Muslim. Cambridge Analytica had been hired by a group of Nigerian billionaires who were worried that if Buhari won the election, he would revoke their oil and mineral exploration rights, decimating a major source of their income.

True to form, Cambridge Analytica focused not on how to promote Goodluck Jonathan's candidacy but on how to destroy Buhari's. The billionaires did not really care who won, so long as the victor understood loud and clear what they were capable of, and what they were willing to do. In December, Cambridge Analytica had hired a woman named Brittany Kaiser to become 'director of business development'. Kaiser had the kind of pedigree that Nix drooled over. In their first meeting, Nix flirted with Kaiser, saying to her, 'Let me get you drunk and steal all your secrets.' She had grown up in a wealthy area outside of Chicago and attended Phillips Academy, an exclusive private school in Massachusetts (alma mater of both Presidents Bush). She went to the University of Edinburgh and afterward got involved in projects in Libya. Once there, she met a barrister named John Jones who represented not only Saif Qaddafi, Muammar Qaddafi's son, but also Julian Assange of WikiLeaks. Jones was a well-respected member of the British bar. Kaiser started consulting for him and, as a result, became acquainted with Assange. She started working at Cambridge Analytica toward the end of 2014, just as I was leaving.

Cambridge Analytica created a two-pronged approach to swaying the Nigerian election. First they would seek out damaging information – *kompromat* – on Buhari. And, second, they would produce a video designed to terrify people from voting for him. Kaiser travelled to Israel, where, according to her, she was introduced to some consultants by her contacts there. According to internal correspondence I saw about the Nigeria project, Cambridge Analytica also engaged former intelligence agents from a handful of countries. It is uncertain who, if anyone, at Cambridge Analytica knowingly procured the services of hackers, but what is clear is that highly sensitive material about political opponents – which may have been hacked or stolen – somehow ended up in the company's possession. By gaining access to opposition email accounts, databases and even private medical records, the firm discovered that Buhari likely had cancer, which was not public knowledge at the time. The use of hacked material was not unique to Nigeria, and Cambridge Analytica also procured *kompromat* on the opposition leader of St Kitts and Nevis, an island nation in the Caribbean.

The hacking of private medical information and emails was disturbing enough, but the propaganda videos Cambridge Analytica produced were much worse. The ads, which were placed on mainstream networks, including Google, were targeted to areas of Nigeria where the population leaned pro-Buhari. A Nigerian surfing the news would encounter an ordinary-looking clickbait ad – a gossipy headline or a photo of a sexy woman. When the person clicked on the link, he or she would be taken to a blank screen with a video box in the middle.

The videos were short – just over a minute long – and they usually started with a voice-over. 'Coming to Nigeria on 15 February, 2015,' intoned a man's voice. 'Dark. Scary. Very uncertain.' 'What would Nigeria look like if sharia were imposed as Buhari has committed to do?' The answer, according to the video, was the most gruesome, horrifying carnage imaginable. Suddenly the video cut to a scene of a man slowly sawing a blunt machete back and forth across a man's throat. As blood spurted from the victim's neck, he was thrown into a ditch to die. The earth around him was stained red. In another scene, a group of men tied up a woman, then drenched her in petrol and set her on fire as she screamed in agony. These were not actors – this was actual footage of torture and murder.

A number of people left CA right after I quit, reasoning that if the firm had become too sketchy for me, the guy who knew all the secrets, then it was too sketchy, *period*. The Nigeria project, a new low, set off another round of departures. By March 2015, everyone I cared about – Jucikas, Clickard, Gettleson and several others – had left Cambridge Analytica. But many others found a reason to stay. Kaiser stayed on until 2018, coming forward publicly after the firm was sinking under the weight of the evidence I had provided to the media and authorities. She later claimed not to have known CA was hiring hackers, telling a British parliamentary inquiry that she just thought they were good at 'intelligence gathering' and using 'different types of data software to trace transfers between bank accounts . . . I don't really know how that works.'

AS I LOOK BACK at my time at Cambridge Analytica, some things make a lot more sense than they did in the moment, when I became conditioned to the weirdness of the place. There were always strange people coming and going – shady characters in dark suits; African leaders wearing oversized military hats the size of dinner platters; Bannon – so if every unusual event tripped you up, you wouldn't have lasted long.

I know now that Lukoil has a formal cooperation agreement with the Russian Federal Security Service (FSB) – the successor to the Soviet KGB. And a member of the House Intelligence Committee later informed me that Lukoil often served as a front for the FSB, conducting intelligence gathering on its behalf. Lukoil executives had also been caught conducting influence operations in other countries, including the Czech Republic. In 2015, Ukrainian security services accused Lukoil of financing pro-Russian insurgencies in Donetsk and Luhansk. 'I have only one task connected with politics, to help the country and the company,' Lukoil's CEO, Vagit Alekperov, said of his role in geopolitics.

In fact, this is likely the primary reason they would have been interested in SCL. SCL had a long history in Eastern Europe, and in 2014 it was in discussions for another NATO project on counter-Russian propaganda. SCL had previously worked on campaigns in the Baltics that blamed Russians for political problems. 'In essence, Russians were blamed for unemployment and other problems affecting the economy,' said one old report on the project. But beyond all that, just as Lukoil was funding pro-Russian insurgencies in Donetsk, SCL's defence division was beginning countermeasure work to 'collect population data, conduct analytics, and deliver a data-driven strategy for the Ukrainian government in pursuit of their goal to win back control of Donetsk.' This project was designed to 'erode and weaken the Donetsk People's Republic (DPR)' and would have made the firm a significant target for Russian intelligence gathering, which was known to operate through Lukoil in Europe.

In reality, when Nix and I met with these 'Lukoil executives,' we were almost certainly speaking to Russian intelligence. They likely were interested in finding out more about this firm that was also working for NATO forces. That's likely also why they wanted to know so much about our American data, and Nix probably struck them as someone who could be flattered into saying pretty much anything. It's entirely possible that Nix did not know to whom he was speaking, just as I did not. What made these contacts all the more concerning was that they wouldn't have needed to hack Cambridge Analytica to access the Facebook data. Nix had told them where it could be accessed: in Russia, with Kogan.

This is not to say that Kogan would have even known about this, but gaining access to the Facebook data would have been as simple as keylogging his computer on one of his lecture trips to Russia. In 2018, after the UK authorities seized Cambridge Analytica's servers, the Information Commissioner's Office subsequently stated that 'some of the systems linked to the investigation were accessed from IP addresses that resolve to Russia and other areas of the CIS.'

It's eye-opening to summarise what was going on over those final months of my tenure. Our research was being seeded with questions about Putin and Russia. The head psychologist who had access to

Facebook data was also working for a Russian-funded project in St Petersburg, giving presentations in Russian and describing Cambridge Analytica's efforts to build a psychological profiling database of American voters. We had Palantir executives coming in and out of the office. We had a major Russian company with ties to the FSB probing for information about our American data assets. We had Nix giving the Russians a presentation about how good we were at spreading fake news and rumours. And then there were the internal memos outlining how Cambridge Analytica was developing new hacking capacity in concert with former Russian intelligence officers.

In the year after Steve Bannon became vice president, Cambridge Analytica started deploying tactics that eerily foreshadowed what was still to come in the 2016 American presidential election. To get access to their opponent's emails, Cambridge Analytica made use of hackers, some of whom may have been Russian, according to internal documents. The hacked emails CA procured were then used to undermine their opponent, including a concerted effort to leak rumours about the opposing candidate's health. And this stolen *kompromat* was then combined with widespread online disinformation targeting social media networks. The overlap in events could be entirely coincidental, but many of the personnel who worked on Nigeria also worked on CA's American operations. One year after Nigeria, Brittany Kaiser was appointed director of operations of the Brexit campaign Leave.EU, and Sam Patten would later go on to work with Paul Manafort on the Trump campaign. In 2018, Patten was indicted by special counsel Robert Mueller and later pled guilty to failing to register as a foreign agent. His business partner, Kilimnik, was also indicted but avoided prosecution by staying in Russia. It wasn't until later, after Patten was revealed to be associated with suspected Russian intelligence operatives, that I wondered again about those bizarre research projects on Vladimir Putin and Crimea.

Patten also ran research in Oregon, which included an extensive amount of questioning about attitudes toward Russian foreign policy and Putin's leadership. Why would Russia care how Oregonians felt about Vladimir Putin? Because once CA modelled people's responses to the questions, the database could identify cohorts of Americans who held pro-Russian views. The Russian government has its own domestic propaganda channels, but one of its global strategies is to cultivate pro-Russia assets in other countries. If you're interested in spreading your narratives digitally, it's helpful to have a roster of people to target who are more likely to support your country's worldview. Using the internet to cultivate local populations with Russian propaganda was an elegant way to bypass all western notions of 'national security.' In most western countries, citizens have free speech rights – including the right to agree with a hostile nation's propaganda. This right serves as a magical force field for online propaganda. US intelligence agencies cannot stop an American citizen from freely expressing political speech, even if the speech was cultivated by a Russian operation. Intelligence agencies can work only on *preemptive* actions to block weaponised narratives from an American social network.

Russia has always been contemptuous of America's approach to free speech and democracy in general. When Russian leaders look at America's history of mass movements and protests, they see nothing but chaos and social disorder. When they look at US courts citing civil rights to permit gay marriage, they see western decadence leading America into weakness and moral decline. To Moscow, civil rights and the First Amendment are the American political system's most glaring vulnerabilities. And so the Russian state sought to exploit this vulnerability – to hack American democracy. It would work, they decided, because American democracy is an inherently flawed system. The Russians created their self-fulfilling prophecy of social chaos by targeting and domesticating their propaganda to American citizens of similar worldviews, who would then click, like, and share. These narratives spread through a system of constitutionally protected free speech, and the US government did nothing to stop them. Neither did Facebook.

Was Cambridge Analytica involved in Russian disinformation efforts in the United States? No one can say for sure, and there's no single 'smoking gun' proving that Cambridge Analytica was the culprit, aided and abetted by Russia. But I've always hated the expression 'smoking gun,' because it means nothing to an actual investigator. Instead, investigators compile small pieces of information – a fingerprint, a saliva sample, tire tracks, a strand of hair. In this case,

Sam Patten worked for CA after working on pro-Russian campaigns in Ukraine; CA tested American attitudes toward Vladimir Putin; SCL's work for NATO made it a Russian intel target; Brittany Kaiser used to consult for Julian Assange's legal team; the head psychologist who was collecting Facebook data for CA was making trips to Russia to present lectures about social media profiling, one of which was titled 'New Methods of Communication as an Effective Political Instrument'; CA systems were accessed by IP addresses that resolved to Russia and other CIS countries; memos referenced ex-Russian security services; and we have Alexander Nix telling Lukoil about Cambridge Analytica's US data sets and disinformation capacity.

When I had lunch with Nix to tell him I was quitting, he was clear about how he thought things would play out. 'The next time you see me,' he said, 'I will be at the White House. And you will be nowhere.' As it turned out, he wasn't that far off. When I next saw Alexander Nix, nearly four years after I told him I was quitting, he was in the British Parliament, answering questions about lies he had told in a parliamentary inquiry. His reputation was being eviscerated before my eyes, but, characteristically, he didn't seem to realise it – or maybe he just didn't care. When he saw me sitting in the gallery, he simply winked.

9

CRIMES AGAINST DEMOCRACY

In January 2016, I decided to accept an offer to consult for the Liberal Caucus Research Bureau (LRB), based at the Canadian Parliament. Justin Trudeau had just formed a new government after leading the Liberal Party to a huge victory in the October 2015 federal election. One of the planks of the Trudeau election platform was reinstating the census, which had been abolished by the previous Conservative government, and reinvigorating Canadian social programs with more data-informed policy making. Just after his victory, I was asked by some of my former Liberal colleagues if I was interested in working on Trudeau's new research and insight team, with a focus on technology and innovation.

After several exceptionally frustrating years, first with the Liberal Democrats in the UK's coalition government and then with Cambridge Analytica, I was desperate to find something to do where I knew I would be contributing something good to the world. It would mean that I would have to return to Canada, but I negotiated an arrangement whereby I would not have to stay in Ottawa, save for important meetings. Having been away from Canada for over five years, I did not have that many friends in the country, but I was reeling from the trauma of everything that had happened, so I thought some calmer downtime back home would help me recover.

When I first arrived in Ottawa for my preliminary meetings and induction, I was flooded with memories of my younger years at

Parliament, trying to get VAN set up. This place had been the setting of my formative adventures working for the leader of the opposition, and now I was back to close a chapter of my life that had started when I was a teenager. Ottawa was the same dull city I had left years before, but, coming from London, it was now even more acutely monotonous. In true Canadian fashion, Ottawa was like an even blander version of Washington, DC – the Diet Coke of capital cities.

The home of the government's political research unit, at 131 Queen Street, was no less bland than the rest of Ottawa, with a vibe somewhere between space station and purgatory. Navigating the building's windowless halls and undecorated beige rooms, I soaked in its bureaucratic aesthetic, occasionally passing reception desks with little blue ENGLISH/FRANÇAIS signs, reminding passersby that in Canada we also *parlons français*. My job description promised steady boredom – basic technical setup, polling advice, social media monitoring, some simple machine learning work and research on innovation policy. Nothing spectacular and, ironically, nothing very innovative came out of it, but I was okay with that, as I wasn't obligated to actually stay in Ottawa. I could quickly flee the LRB office in Ottawa and work on projects around Canada, which would keep me sane.

Meanwhile, back in Britain, Conservative Prime Minister David Cameron had announced a referendum on the country's future – whether it would continue to be a member of the European Union or strike out on its own. Ever since the UK joined the European Economic Community (EEC) in 1972, Eurosceptics had been agitating for withdrawal. Initially, the left wing led the movement, with many Labour politicians and trade unionists agreeing that a bloc-style pact would harm their socialist dreams. But most of their countrymen welcomed the arrangement. When a referendum in 1975 asked the British public if they wanted to stay within the European Economic Community, the vote was 67 per cent in favour.

By the time the EEC became the European Union, the left and the right were largely in agreement on the benefits of membership. But in the early 1990s, the right-wing UK Independence Party (UKIP) emerged out of a growing resistance to European priorities. In 1997, Nigel Farage, a former commodities trader and founding UKIP member, ousted the leader of the party. Farage became leader in 2006, and under his leadership, UKIP began stoking virulent anti-immigration sentiment among working-class whites and tapping into nostalgia for Britain's imperial past in wealthy white communities. The world had been transformed by the 11 September attacks, the rise of Islamophobia, and the conflicts of the Bush and Blair years. As the fate of black and brown refugees developed into a European crisis, Cameron moved to appease nationalist sentiment to retain right-wing voters. The Conservative Party drew up a plan for a referendum, to be held before the end of 2017. The date was set for 23 June, 2016.

REFERENDUMS IN BRITAIN ARE largely publicly financed, where each side of the ballot question receives equal amounts of public funding after the UK's Electoral Commission designates one campaign group on each side to become the official campaign. British electoral law also sets strict spending limits, applied equally to both sides, to ensure that one side is not unfairly advantaged by more money than the other. In effect, these are Britain's electoral equivalent of Olympic anti-doping rules that ensure a fair race. Having more resources means being able to reach a disproportionate number of voters with one's messaging, so the resources are regulated to maintain a fair election. Other groups are still allowed to campaign, but they do not receive public funding and they may not coordinate their campaigns without declaring the spending against the official limit.

Politicians and campaigners had until 13 April, 2016, to win designation as the official campaign for Leave or Remain. Vote Leave and Leave.EU were among the main Leave campaigns. Britain Stronger in Europe was the official campaign for Remain from the start, with specialist initiatives such as 'Scientists for EU' and 'Conservatives In' also campaigning to remain in the union. Vote Leave consisted largely of Conservatives, with a handful of Eurosceptic progressives. The other pro-Brexit campaign, Leave.EU, was focused almost entirely on immigration, with many of its campaigners peddling racist tropes and far-right talking points in order to rile up the public. Each group had its own targets and ideological strategies, and, according to British law, they could not work together in any fashion. Eventually, Vote Leave and Britain Stronger in Europe were granted official campaign status by the Electoral Commission. But the two main Leave groups set themselves up to push different buttons among potential supporters – a tactic that was spectacularly effective in generating votes.

College-educated city dwellers, accustomed to living among immigrants and working in businesses that benefit from their skilled labour, rejected right-wing fearmongering and generally supported Remain. Lower-income Britons and those who lived in rural areas or old industrial heartlands were much more likely to support Leave. National sovereignty has always been a core part of British identity, and the Leave campaign argued that EU membership was undermining that sovereignty. Remain supporters countered by pointing to economic, trade, and national security benefits in the status quo.

Vote Leave was led in public by the campaign's lead spokesperson, Boris Johnson, a pompous man who was once mayor of London and destined to become prime minister, was always a Conservative favourite, with some of the highest approval ratings among Conservative voters, and Michael Gove, who could be characterised as Johnson's opposite. Lacking Johnson's pomposity, Gove was more measured and was a favourite among the free-market-type libertarians in the UK Their slogan, 'Vote Leave, Take Back Control,' was laughed at by Remain camps, but it was not really about the EU itself. It was meant to appeal to voters who otherwise felt their lives were not in their control – their lack of job prospects or an education meant that their lives, more than anyone else's, were more susceptible to the winds of a bad economy and a British society that systemically ignores them. Vote Leave had been co-founded in 2015 by Dominic Cummings, one of Westminster's most infamous political strategists, and Matthew Elliott, founder of several right-wing lobbying groups in the UK. Some in the Vote Leave office disagreed on politics, but they were united under Cummings's leadership behind the scenes.

While Vote Leave operated from the seventh floor of Westminster Tower, on the banks of the River Thames, directly across from Parliament, Leave.EU was based more than a hundred miles away, in Lysander House, Bristol, overlooking a busy roundabout. The group shared an office building with Eldon Insurance, a firm run by millionaire Arron Banks, who also happened to be the co-founder and main funder of

Leave.EU. The campaign launched during the summer of 2015 and teamed up with Cambridge Analytica in October of that year. The Eurosceptic Nigel Farage, a prominent right-wing politician, became the figurehead for Leave.EU. After Steve Bannon introduced Banks and Farage to the American billionaire Robert Mercer, Cambridge Analytica signed on to the Brexit campaign to service Leave.EU with its algorithms and digital targeting. It was announced that Brittany Kaiser would become Leave.EU's new director of operations, with Kaiser and Banks launching Leave.EU together at a press conference.

Shortly before returning to Canada, I had drinks with a few of the people I'd gotten to know during my time in British politics. One of those was a special adviser to then-Home Secretary Theresa May, a gay Conservative named Stephen Parkinson. He was a Tory, but something I had learned over the years in politics is that it's usually easier to be friends with people outside of your own party, because they aren't in direct competition for your job and are less likely to try to screw you over personally. Parkinson told me he had just taken a leave of absence from the Home Office to work for Vote Leave – a newly created campaign group for Brexit. I wasn't surprised that Parkinson was working on it, and I told him I knew a few other people who might be interested in joining his campaign.

One was a young student at the University of Brighton named Darren Grimes. I had originally met Grimes through the Liberal Democrats, but he had become disenchanted when the party began to implode in the internal leadership race that followed its decimation in the 2015 elections. When Grimes decided to leave the Lib Dems, he asked me for an introduction to the Tories, so I introduced him to Parkinson. You've probably never heard of Grimes, but he would later become an accidental central player in Vote Leave's victory in the Brexit referendum.

Parkinson and I met several times before I left London, because he wanted to get my thoughts on data analytics. He did not say this at the time, but he knew about Cambridge Analytica and clocked how valuable such targeting tools would be for the Brexit campaign. He said that he wanted to introduce me to someone. 'His name is Dom Cummings.' I flinched at the name.

Dom Cummings – not a porn name, though it would be a good one

– had made a reputation at the Department for Education in the coalition government as a Machiavellian operator and a very difficult character. Cameron, the prime minister at the time, would later suggest that Cummings was a 'career psychopath'. In keeping with his notoriety, Cummings went on to become the mastermind behind the largest breach of campaign finance law in British history, using some of the technologies developed at Cambridge Analytica to tilt the Brexit vote towards Leave. But I didn't learn any of this until it was too late: at the time, he was just an abrasive, ambitious Conservative staffer who enjoyed irritating the shit out of everyone in the British political system.

Parkinson, Cummings and I all sat down in a barren room in the future headquarters of Vote Leave to talk about voter targeting. The entire floor was being renovated and was covered in plastic sheeting, but, sitting along Albert Embankment, it had spectacular views of the Palace of Westminster, directly across the Thames. My first impression of Cummings was that he looked dishevelled, as if he'd just climbed onto a lifeboat after the *Titanic* sank. Cummings has a very large head, and his hair tends to go all over the place, with wispy strands haphazardly crossing his balding pate. He looked a bit dazed, or maybe a bit blazed, like he was either stuck trying to solve a puzzle or had just dragged an epic joint – I could never quite tell.

To his credit, Cummings is one of the few smart people I have encountered working in the Augean stable of mediocrity that is British politics. What I liked about meeting Cummings was that we didn't talk about what people in politics usually obsess about. Cummings understood that more people are busy watching the Kardashians or Pornhub than following the political scandal *du jour* on BBC *Newsnight*. Instead, Cummings wanted to talk about identity, about psychology, about history, and, indeed, about AI. And then he mentioned Renaissance Technologies, the hedge fund set up by Robert Mercer. Cummings had obviously read up on Cambridge Analytica, and he asked a lot of questions about how the firm worked. He was interested in creating what he called 'the Palantir of politics' – a term I shuddered at after hearing it used so often by Nix. I just rolled my eyes, thinking, *Here we go again*.

Vote Leave did not even have the electoral register yet, so I told Cummings that I was extremely sceptical that he could develop data sets anywhere close to those used by Cambridge Analytica. And I continued to tell him that Steve Bannon was close with Nigel Farage, so the chances were high that Cambridge Analytica was already working with their rival pro-Brexit campaign, Leave.EU. Shortly after the meeting, Leave.EU officially announced its partnership with Cambridge Analytica, apparently spoiling Cummings's plan. After the meeting, Parkinson invited Gettleson and me to come work for Vote Leave. As I had already accepted a project working for Justin Trudeau, I declined. But Gettleson, after initially flirting with the idea of joining me in Canada, decided to stay in London and work for Vote Leave, because he was not in a place for another dramatic change in his life like moving to a new country. As a courtesy, I nonetheless sent Cummings an email outlining how he could probably attempt a pilot of *a few thousand* voter surveys, but I estimated that that was about all they could accomplish in the exceptionally tight time frame of the referendum – well, at least all they could do *legally*.

Just before I left for Ottawa, another friend of mine in London named Shahmir Sanni asked if I could help him find an internship. He and I had first connected on nights out in London, and we kept in touch over Facebook, frequently trading thoughts and opinions on politics, fashion, art, hot boys and culture. Sanni had just finished university and was interested in politics, but he had no connections and needed an introduction. I asked him where he wanted to join, but he said party was not a concern; he was most interested in gaining experience. When I asked contacts in both the Remain and Leave campaigns about internships, only one responded: Stephen Parkinson. Parkinson asked who I wanted to introduce, so I texted him Sanni's Instagram profile. Parkinson, clearly enamoured with Sanni's well-curated photos, texted back just two words: 'YES PLEASE!!!!!' And that's how Sanni – who would eventually become one of two Brexit whistleblowers – signed on to the Vote Leave campaign.

Pro-Brexit leaders knew that they weren't going to win the vote by speaking only to traditional right-wing Brexiteers, so Vote Leave made it a priority to bring in a more diverse coalition of support. In British politics, referendum campaigns are unique in that they tend to make a concerted effort to be as cross-party as possible, because issues, not parties, are on the ballot. No one 'wins power' at the end of a referendum; only the idea wins, and the government of the day has a choice whether or not to implement the result. Cummings and Parkinson understood that the key to a Brexit victory was to identify Labour and Lib Dem voters, as well as those who didn't normally vote, and persuade them to either vote Leave or stay neutral. It was for this reason that the pro-Brexit side was extremely eager to recruit Lib Dems, Greens, Labour, LGBTQ, immigrants – as many traditionally non-Conservative voters as possible. Sanni was the perfect man to help with that mission.

One of the most compelling progressive arguments for Brexit was pretty simple. It was that the European Union tended to favour European – i.e., white – immigrants over those from the Commonwealth nations, who were predominantly people of colour. Under EU rules, migrants to Britain from countries like France, Italy, Spain, Germany and Austria did not need a visa to work and live in Britain. But migrants from, say, India, Pakistan, Nigeria or Jamaica were required to undergo extensive screening and difficult immigration procedures. Yet for hundreds of years, Britain had built its vast empire predominantly by using the labour of people of colour throughout the Commonwealth, conquering their lands, taking their resources, and leaving them to struggle at home while the great cities of Britain flourished in the wealth created from abroad. In both world wars, when British freedom was threatened by other European nations, the citizens of the Commonwealth were called to arms to fight for Britain. Few, if any, major war films have been made to honour their sacrifice, but many of the great British victories were in fact won with the spilled blood of Commonwealth soldiers from India, the Caribbean and Africa. Then, decades later, when Europe looked more economically promising than the fledgling countries emerging out of colonial rule, Britain turned its back on these nations, closed off its borders, and implemented tough new immigration rules for Commonwealth citizens. At the same time, Britain began opening up nearly unrestricted immigration to European citizens, who were overwhelmingly white.

It was out of this sense of deep unfairness that many people of colour – people like Sanni's friends and family, who were from Pakistan – had no affinity for the EU: they knew what it felt like to have to endure a Kafkaesque immigration system requiring them to prove every ounce of their worth. They knew what it felt like to live in a country that had exploited their ancestors to build itself up but now sent Home Office trucks roving through the neighbourhoods of Indian and Pakistani communities, emblazoned with warnings like HERE ILLEGALLY? GO HOME OR FACE ARREST. TEXT HOME TO 78070. Meanwhile, a German or Italian, whose grandfather may very well have taken deadly aim on those battalions of Indians and Nigerians that Britain called into battle, could enter Britain with no questions asked, and then get busy applying for jobs.

As the Remain campaign paraded around its 'pro-immigration' messages to defend the EU, what many people of colour saw was the tacit whiteness of that very message – that it really meant rights for *some immigrants*. For people like Sanni, Brexit was a story of marginalisation and of Britain's unaddressed legacy of colonialism – an attempt to right the wrong of denying immigrants and people of colour access to the very country that had plundered them for centuries. And it was by identifying this bubbling resentment that the pro-Brexit movement managed to create a counterintuitive alliance between some sections of immigrant communities and cohorts of jingoist Brexiteers who wanted them all to 'go home'.

PARKINSON GAVE SANNI AN unpaid internship. He started in the spring of 2016 as a volunteer. Because the outreach team was so small, his duties quickly multiplied. Much of his work was focused on minority and queer communities. He would visit impoverished neighbourhoods to ask residents how they were planning to vote and why.

On Sanni's first day at the office, he noticed a dandy in a green blazer and pink trousers: Mark Gettleson, in his full homosexual plumage. Right away, Sanni and Gettleson started joking about being the odd men out in this sea of conservative white men. Gettleson had joined Vote Leave as a consultant in the spring of 2016, impressing the staff with his wit, intelligence and intuitive understanding of British liberals. He immediately began setting up the websites for several of the outreach groups, many of which he branded and named himself – Green Leaves, Out and Proud and others. When Darren Grimes, the twenty-two-year-old fashion student I knew from the Lib Dems, joined the team, he and Gettleson started conceptualising a progressive arm of Vote Leave, to be called BeLeave.

I was in Canada by then, but we all kept in touch via Facebook. In the course of designing the branding for BeLeave, Grimes sent me his idea sheet via Messenger. Even though I was preoccupied with setting up projects for the new Liberal government in Ottawa, I wanted to give him a hand after the rough time he'd had with the Lib Dems. One of his challenges was in choosing the right colours. The official colour for Vote Leave was red, so they needed something different. I said, 'Why not use the Pantone colours of the year?' – which in 2016 happened to be Serenity blue and Rose Quartz pink. Darren did a mock-up, and I messaged back, 'It looks so gay and millennial. Not fascist at all.'

BeLeave attempted to appeal to the softer side of the pro-Brexit vote by focusing on issues such as parity in treatment for immigrants, ending what they termed 'passport discrimination' between EU and non-EU citizens, the unfair impact protectionist EU policies had on African farmers, and environmental protections. After Parkinson asked Sanni to shift his attention from minority outreach to BeLeave, he and Grimes – a pair of interns in their early twenties – essentially ran the initiative, with occasional input from the senior staff at Vote Leave. With the hardcore anti-immigration votes already in the bag, the Leave side needed to secure only a small percentage of more liberal-minded voters to win. Data was the key to targeting those voters.

Vote Leave didn't have the data it needed, though, and the one company that could provide it, Cambridge Analytica, was not an option because it was already working with Leave.EU. If Vote Leave worked with Cambridge Analytica, it would run afoul of laws restricting coordination between the campaigns. What they did, I learned later, was hire a firm whose origins intersected with my early days at SCL, when I was just starting to put together a technical team.

This was back in August 2013, when I was looking for people who could help. I recalled my time at the LPC and the mentorship of Jeff Silvester, who had taken an interest in me while I was still in school. Silvester, a computer software engineer by training, had developed a solid understanding of enterprise data systems long before he started advocating for a new data strategy at the LPC. A big guy with a beard

– he reminded me of Ron Swanson in *Parks and Recreation* – Silvester was unfailingly considerate and thoughtful, but he also had the dry and cynical sense of humour of someone who had spent years in politics. He lived outside of Victoria, British Columbia, and on weekends he helped mentor young people as leader of a local Boy Scout troop. My first few months working with Silvester as an intern involved helping him on casework for refugee and political asylum claims, and he showed me how you can really make a difference in people's lives. He was one of the most honorable people I knew.

Shortly after joining SCL, I wrote to Silvester, describing the firm's portfolio – not just psychological warfare projects for NATO, but efforts to fight HIV in Africa. He quickly replied: 'You need a Canadian office!' When the Trinidad project came up, he got his wish. SCL needed someone to help build and manage data infrastructure, and Silvester had exactly the right background. Silvester poached another Canadian political operative, Zack Massingham, a veteran of the rough world of BC provincial politics, to lead project management for the new company, which he called AIQ. The company was registered in Canada and was legally called AggregateIQ, but it signed an intellectual property agreement that granted SCL the rights to its work. SCL and, later, Cambridge Analytica frequently took advantage of a network of offshore companies registered under different names. Similar to the strategies employed by tax avoidance schemes, this network of companies around the world helped Cambridge Analytica bypass the scrutiny of electoral or data privacy regulators.

AIQ's headquarters was a brick building on Pandora Avenue, only a block from the ocean in Victoria, on Vancouver Island. SCL and CA employees loved to visit the office – it was scenic, breezy and relaxed compared with the frenetic pace of London. As AIQ grew, the firm recruited a fantastic and diverse team of engineers to work on SCL projects.

AIQ's Trinidad contract with SCL included building infrastructure for Facebook data harvesting, clickstream data, ISP logs and the reconciliation of IPs and user agents to home addresses, which would help de-anonymise internet browsing data. As SCL grew into Cambridge Analytica, AIQ became an indispensable part of the back-end technical engineering team. Once it was decided that CA's models would have to be loaded into a platform that could launch social and digital ad targeting, AIQ was tasked with constructing Ripon, CA's ad-targeting platform. After Kogan harvested the Facebook data, it was passed on to AIQ for loading into the Ripon platform, which allowed a user to segment universes of voters according to hundreds of different psychometric and behavioural factors. During the 2016 US primaries, AIQ staff members would travel south to Texas to build out infrastructure for Senator Ted Cruz's campaign.

When Brittany Kaiser and Sam Patten joined Cambridge Analytica and took over the Nigeria project, AIQ was brought on to distribute CA's voter suppression and intimidation propaganda. After uploading videos of women being burned alive and men choking on their own blood as they had their throats cut, AIQ sought to target the content at regions and voter profiles that CA gave them. In 2015, when I found out that Silvester was working on this project, it felt truly bizarre: my onetime mentor was not at all the sort of person who would blithely disseminate videos of torture victims. Years later, I would meet up with Silvester and ask him about Nigeria. Unless you count uncomfortable laughter, he showed no remorse. Somehow he had made peace with the mayhem his company had spewed out into the world as a Cambridge Analytica contractor.

ON THE AFTERNOON OF 16 June, 2016, a pro-Remain Labour MP named Jo Cox was walking to the library in the small town of Birstall, West Yorkshire. She was heading to her fortnightly surgery with constituents who need help with casework or want to raise an issue. But when Cox was just steps from the library door, a man wearing a baseball cap walked toward her, raised a sawn-off shotgun, shouted 'Britain first!' and shot her point-blank. The man then dragged the forty-one-year-old Cox between two parked cars and began stabbing her, flailing his knife at shocked onlookers who tried to stop him. He continued to yell 'Britain first! This is for Britain!' throughout the attack, which finally ended when he reloaded his gun and shot Cox in the head. Cox, the mother of two young children, lay dying on the pavement.

The assassination of Jo Cox sent waves of genuine shock throughout Britain, where gun violence is far less common than in the United

States. Fellow MPs gathered for a vigil on Parliament Square, where flowers left by mourners formed a makeshift memorial. It soon emerged that the killer was a white supremacist and Nazi sympathiser, which only served to heighten the emotional tension between Leave and Remain supporters. In an attempt to calm the storm, and in tribute to Cox, the Leave and Remain campaigns agreed to stop all campaign activities for three days, an extraordinary decision with only a week remaining until the vote. However, AIQ secretly continued deploying digital advertisements for Vote Leave, knowing that the British media would not be able to tell whether they were continuing online ads. It seemed that, after distributing videos in Nigeria of people being tortured and killed, a bit of extra digital campaigning during a period of public mourning for a murdered MP was not beneath them.

By this time, the political climate in Britain had become extremely toxic. Threats were being sent to both Remain- and Leave-supporting MPs (mostly to the Remain side), there was a disproportionate increase in race-based violence, and social media was blowing up every day. No one was passive or nonchalant about what was going on in British politics anymore. People were awake and people were angry. Very angry.

A lot of the messaging from the Leave side during this time was targeted toward 'metropolitan elites', as the politicians called them, as well as people of colour and European migrants. Vote Leave eschewed responsibility, but it was apparent that they had left the race-baiting to Leave.EU, which gladly (and proudly) took up the cause. A few days before Jo Cox was murdered, Leave.EU's Farage unveiled a campaign poster showing a caravan of brown-skinned migrants beneath the words BREAKING POINT. The move drew comparisons to Nazi propaganda from the 1930s showing lines of Jewish people flooding into Europe.

As I sat in Canada watching the drama unfold, I told myself that Vote Leave was not the same as Leave.EU, as many of my friends were working for Vote Leave. *Farage's campaign is the racist one using Cambridge Analytica,* I thought. *Vote Leave couldn't possibly be pandering to that kind of rhetoric.* I was wrong.

By the final weeks of the campaign, Vote Leave had spent nearly all of its allotted £7 million. British law barred it from accepting any more funds or collaborating with other campaigns, but Cummings wanted to keep spending and decided to find another way. AIQ had been receiving the bulk of Vote Leave's ad spending, and Cummings was extremely impressed with the power of AIQ's digital targeting capabilities. AIQ was able to target, engage and enrage specific voters. Many of AIQ's targets were infrequent voters, so even though public polls had Remain ahead, this meant AIQ was engaging new niches of the electorate who were systemically excluded from traditional campaigning and by polling firms. But AIQ realised that if it was to sustain its momentum, it would need more money than Vote Leave was legally allowed to spend – and it needed it fast. So attention turned to the BeLeave project. Until that point, BeLeave had been a totally organic operation run by a couple of interns in the Vote Leave office. There were no paid ads, and all the creative content was being developed by Sanni and Grimes in their spare time. Vote Leave would provide guidance and money for certain things, but only very small amounts – £100 here and there.

Around the same time, Parkinson began inviting Sanni to stay after hours with him in his house because he knew that Sanni was based in Birmingham, and the two began a relationship. For Sanni, who was twenty-two and not yet out about his sexuality to his family, this was all very new and confusing. He did not know how to deal with this intimate situation he now found himself in with his boss. But he was in awe of receiving so much attention and guidance from a senior political adviser who worked at the top echelons of the British government. Parkinson would take Sanni out and tell him how pleased he was with his work, and that if he kept it up there might be a future career in it for him. Sanni agreed to keep the affair secret.

Voters were noticing BeLeave's work too. Some of the content that Sanni and Grimes had created went viral and even outperformed Vote Leave's paid ads. BeLeave's graphics focused on progressive issues such as the 'tampon tax', arguing that if Britons were out of the EU, they would not need twenty-seven other member states to agree to get rid of what was so obviously a misogynistic tax. It seemed that there was a clear market for BeLeave's progressive, woke and social-justice-oriented brand of Euroscepticism. Weeks before the 23 June referendum, Cleo Watson, who was head of outreach in Vote Leave, set up a meeting for Grimes and Sanni with a potential donor. The pair met the donor in the Vote Leave headquarters and made a proposal outlining how effective their posts were – their organic outreach was in some cases surpassing the impact of Vote Leave's paid ads.

Grimes had sent me this presentation, asking my advice on how to optimise targeting online for Facebook and what his budget should be. I messaged him back, providing guidance on what metrics they should use and how they should pitch. It was a good presentation, but the donor ultimately decided against putting in any money. After the donor pulled out, one of Vote Leave's senior directors approached the two young interns and told them that they had found a new way to get them money for BeLeave – but they would have to sign some paperwork first. After meeting with Vote Leave's lawyers, Sanni and Grimes were instructed to set up as a separate campaign, open a bank account, and write a formal constitution. The Vote Leave lawyers drafted the new campaign's articles of association and handed the interns paperwork to sign. What Sanni and Grimes did not realise was that it was not lawful for BeLeave to spend any more money because it was working so closely with Vote Leave. By saying that the BeLeave campaign was separate and could spend its own budget Vote Leave was putting these young interns at risk for any illegal campaign spending undertaken by this 'separate' campaign. But the two interns were not told any of this, and they continued as before, working at the Vote Leave headquarters, attending Vote Leave events, and helping with leafleting.

The next week, Grimes and Sanni were told that the money Vote Leave promised them was finally coming through – and it was going to be more than they had asked for. In fact, it was going to be *hundreds of thousands of pounds* more. Vote Leave began organising the transfer of £700,000 to BeLeave, in what would be the single largest expenditure of Vote Leave's entire campaign. But Grimes and Sanni had to first agree to one condition. The problem for Vote Leave was that if the two were to receive the money as an 'independent' campaign, they would be legally entitled to spend it however they liked. So Vote Leave told the two interns they would never actually see any of the money in their new bank account. Rather, Vote Leave would transfer the money to AIQ directly, and Grimes and Sanni would simply have to sign off on a set of AIQ invoices. Disappointed, Sanni asked if he could at least have his travel and food costs covered by some of these funds (he was treasurer and secretary), but he was told by his Vote Leave supervisor that that would not be possible. Grimes and Sanni had no idea that what they'd just agreed to was completely illegal. They had trusted the lawyers and advisers of Vote Leave, who consistently told them everything was in order.

What made that deception even worse was that Vote Leave's lawyers put these interns' names on the BeLeave documents, making Grimes personally liable for the legal fallout that eventually came. This was not an uncommon strategy in some of the dirtier schools of British campaigning, particularly among the Tories, who have been caught several times using the scheme: senior campaign advisers not wanting to take on the personal risk of breaking election laws would find someone inexperienced, often an eager young volunteer, and nominate them as the campaign's 'agent', which would make that person legally liable for the campaign. That way, if and when wrong-doing was uncovered, a fall guy was in place and the true perpetrators could walk off scot-free, continuing to enjoy their proximity to power while leaving behind the betrayed volunteers and broken lives.

FINALLY, THE DAY OF the referendum arrived. On 23 June, torrential rains continued to batter the south of England, and Londoners hurrying to vote ended up facing horrendous delays, with train stations closing and the tube shutting down in the evening due to flooding. Most of the Vote Leave team, including Grimes and Sanni, spent the day going to key Leave seats to get out the vote. Dover was the gateway to Europe for the UK by sea and by train, and the final stop for Britons before they entered the English Channel. The volunteers spent many hours in Dover knocking on doors in heavy, torrential rain. The front page of the right-wing tabloid *The Sun* had a single bold and familiar headline: BELEAVE IN BRITAIN.

I had no idea AIQ was involved in the Leave campaign until the night of the vote, when Parkinson texted me a photo of himself with Massingham inside the Vote Leave headquarters – grinning in front of fogged-up windows with the outline of Parliament lingering behind them. Weirdly, even though I had seen and spoken to Silvester several times since returning to Canada, he never *once* mentioned AIQ's connection to the Leave campaigns. After the spending returns were released, it was revealed that AIQ had received 40 per cent of Vote Leave's budget – and hundreds of thousands of pounds more from the other pro-Brexit campaigns, including BeLeave.

Now I understood that this was how Cummings could have gotten around the fact that Cambridge Analytica was already working with Leave.EU – he just used one of CA's subsidiaries, based in a different country, with a name that no one knew. AIQ had Cambridge Analytica's infrastructure, handled all of its data, and could perform all the same functions, but without the label. (Vote Leave denies that it had access to Cambridge Analytica's Facebook data.) Nobody had wanted to tell me because everyone knew that I had left Cambridge Analytica on such bad terms, as had many others. Silvester and Massingham chose to stay quiet because this was their biggest gig in politics. Silvester was perfectly comfortable talking about the shady work they had done in Africa or the Caribbean, but not about Brexit.

As someone who had worked on targeted campaigns, I knew that most of the content that the media was talking about was not what individuals and groups were actually seeing during the referendum. Almost instantly, I realised something deeply sinister was happening in Britain. Even so, 72 per cent of voters cast ballots. For hours, the vote was too close to call, but in the end, Leave emerged victorious with 51.89 per cent of the vote. Unbeknownst to me at the time, Vote Leave had appointed Thomas Borwick to become the chief technology officer of the campaign. Before joining Vote Leave, Borwick had worked with Alexander Nix and SCL running a number of data-harvesting projects in island nations across the Caribbean. (However, there is nothing to suggest that Borwick participated in any of SCL's unlawful work in the region.) After the referendum, Borwick revealed that Vote Leave and AIQ had together disseminated more than a hundred different ads with 1,433 different messages to their target voters in the weeks leading up to the referendum. Cummings later revealed that these ads were viewed more than 169 million times, but only targeted at a narrow segment of a few million voters, which resulted in their newsfeeds being dominated by Vote Leave messaging.

The people of the United Kingdom were the targets of a scaled information operation deployed by AIQ, and the problem with Remain was that they completely failed to understand what they were up against. As Cambridge Analytica identified, provoking anger and indignation reduced the need for full rational explanations and would put voters into a more indiscriminately punitive mindset. CA found that not only did this anger immunise target voters to the notion that the economy would suffer, but some people would *support* the economy suffering if it meant that out-groups like metropolitan liberals or immigrants would suffer in the process – that, in effect, their vote would be used as a form of punishment.

This approach proved effective against Remain's 'Project Fear' messaging, which tried to focus voters on the potentially catastrophic economic risks of exiting the European Union. In short, it is far harder to make angry people fearful. The 'affect bias' arising out of anger mediates people's estimation of negative outcomes, which is why angry people are more inclined to engage in risky behaviour – the same is true whether they are voting or starting a bar fight. If you have ever been in a bar fight, you know that literally the *worst way imaginable* to make your opponent think twice about a rash move is to yell threats at him. It only eggs him on.

Remain's focus on the economy also neglected to stop and ask people what they thought the economy was in the first place. Cambridge Analytica identified that many people in non-urban regions or in lower socioeconomic strata often externalised the notion of 'the economy' to something that only the wealthy and metropolitan participated in. 'The economy' was not their job in a local store; it was something that bankers did. This is also what made certain groups comfortable with economic risks and even trade wars, since, in their minds, that chaos would be unleashed upon the people who worked in 'the economy'. And the more forceful the economic argument they heard, the more confident they would become that what they were 'actually' hearing was the fears of a cowering elite worried about losing its wealth. This made them feel powerful, and it would become a power they wanted to wield.

After Leave won, a wave of shock and consternation swept Britain and the world. David Cameron gave a sombre statement in front of 10

Downing Street, saying he would step down as prime minister by October. Both the euro and the British pound plummeted in value and global stock markets nosedived. A petition began circulating, asking for a second referendum, and within seventy-two hours of the election, more than three-and-a-half million people had signed it. In the United States, the response was mostly surprise and confusion. As pundits tried to parse what Brexit would mean for Americans, President Obama adopted the Keep Calm and Carry On approach, assuring everyone that 'one thing that will not change is the special relationship that exists between our two nations.'

Donald Trump, then the presumptive Republican nominee, happened to be in Scotland at the time, visiting his Trump Turnberry golf resort. He called the Leave victory 'a great thing', saying that the voters had taken back their country.

'People want to take their country back, they want to have independence,' Trump said. 'People are angry, all over the world ... They're angry over borders, they're angry over people coming into the country and taking over, nobody even knows who they are. They're angry about many, many things.'

The world did not know it yet, but Brexit was a crime scene. Britain was the first victim of an operation Bannon had set in motion years before. The so-called 'patriots' of the Brexit movement, with their loud calls to rescue British law and sovereignty from the grips of the faceless European Union, decided to win a vote by mocking those very laws. And to do so, they deployed a web of companies associated with Cambridge Analytica in foreign jurisdictions, away from the scrutiny of the agencies charged with protecting the integrity of our democracies. Foreshadowing what was to come in America, a clear pattern emerged during the Brexit debacle, where previously unknown foreign entities began exerting influence on domestic elections by deploying large data sets of unexplained origins. And with social media companies not performing any checks on the advertising campaigns spreading throughout their platforms, there was no one standing guard to stop hostile entities seeking to sow chaos and disrupt our democracies.

10

THE APPRENTICE

'I'M NOT GOING TO LIE, THIS IS CERTAINLY ONE OF THE WEIRDER cases I've dealt with,' my lawyer said as we sat in her office in London reading through a June 2015 pre-action letter from Cambridge Analytica (falsely) claiming that I was attempting to set up a rival firm in aid of the nascent Trump presidential campaign. Donald Trump had first come into my life a few months before, in the spring of 2015, when Mark Block called me with a proposition refreshingly far removed from the work I'd done with Cambridge Analytica. As Block explained it, the Trump Organization needed help with market research, either for Trump's reality TV show *The Apprentice* or his casinos. Block had also called Jucikas and Gettleson, who were still in London. The three of us talked, and we agreed to a meeting with the Trump executives.

In calls with the Trump Organization, we heard about declining ratings for *The Apprentice* and how fewer people were staying at Trump hotels and gambling in the casinos. With the advent of online gambling and the total dependence on Donald Trump's public image as a sexy, savvy billionaire, it seemed his team was beginning to realise that an outdated casino system and an aging, orange-stained C-list celebrity didn't conjure 'sexy and fun' for potential new customers. The Trump brand was on a downturn, and the company needed to figure out how to give it a boost.

The project was frustratingly amorphous, the executives weren't even sure what we did or how we could help, and I started to suspect that the Trump executives were just looking for free advice. When they proposed a meeting for about a month later, I declined, content to let Jucikas and Gettleson report back from Trump Tower. The meeting took place at a Trump Tower restaurant, and the conversation picked up in the same vague place the calls had left off. Could we use data to enhance the image of Trump and his products – to revive the Trump brand? And if so, who would be the targets for such a project?

When Gettleson called to tell me about what went down, he was laughing. 'You're not going to believe this,' he said. 'Trump is planning to run for president.' The meeting had included Corey Lewandowski, who identified himself as Trump's campaign manager and assured Gettleson and Jucikas that Trump was indeed serious about running for president. He invited us to take part in the campaign – an offer I wanted no part of, for several reasons. One: it was a political campaign, and I had quit Cambridge Analytica and left London specifically to get away from politics. Two: Trump seemed like an utterly ridiculous individual and a likely failure as a candidate. And three: he was running as a Republican, and I was finished doing dirty work for right-wing politicians. It was one thing to explore how to improve ratings for a reality TV show; it was quite another to help a Republican run for president. Gettleson agreed – Jucikas, soon to be a consultant on Republican campaigns, not so much – and the Trump subplot, we assumed, had come to an end.

But a couple of weeks later, on 5 June, 2015, we learned that Cambridge Analytica was suing Gettleson, Jucikas and me. They claimed that we had violated the non-solicitation clause in our NDAs with the firm. We had, according to the lawsuit, solicited one of Cambridge Analytica's clients: Donald Trump. The letters informing us of the lawsuit gave us two weeks to respond, so even though the case was clearly bogus, I decided to hire a lawyer to make it go away as quickly as possible. At our first meeting, the attorneys were baffled. Imagine how strange this conversation was, long before Cambridge Analytica or Steve Bannon became household names: 'So there's this psychological warfare firm,' I told them. 'And it got acquired by this Republican billionaire in the United States. And after I quit, I got invited to talk with Donald Trump – the guy from *The Apprentice*? Apparently, he's going to run for president, and he's secretly a client of theirs. And so now they're suing me . . .'

By now, Cambridge Analytica had spread like a disease through the Republican Party, advising prominent candidates in House and Senate races and undertaking projects to study cultural phenomena, such as militarism among US youth, on behalf of right-wing interests. On the face of it, Cambridge Analytica was wildly successful. But behind the scenes, the firm was screwing the Republican Party generally – and the Mercers specifically. For me, the real revelation in the lawsuit was that Cambridge Analytica was connected with Trump as an off-book client at the same time the firm was working for the Mercers' preferred presidential candidate, Ted Cruz. Not only did Bannon have a different agenda than the Mercers, he also had no interest in supporting Cruz, whom Bannon despised.

After I explained to the attorneys that I wasn't even working for Trump, they essentially said, 'Fine, don't worry. Firms send these letters all the time as a stern warning, but usually nothing comes of them. It's probably their director's insecurity. We can take care of this.'

But it wasn't going to be cheap. Indeed, Cambridge Analytica made it clear that they intended to keep badgering me, costing me money and peace of mind, until I gave in. I offered to sign a document saying I would never again work for a Republican, but Cambridge Analytica didn't want that. They wanted me to never work with data again, which of course was an impossible request. Back and forth we went, for months on end. The whole thing got more and more bizarre. In the course of the legal dispute, I discovered that after Gettleson and I left the firm, CA had invented two fake staff personas – 'Chris Young' and 'Mark Nettles' – that they continued to use on their site and with clients. Finally I agreed to sign a deed of confidence, which was essentially a super-non-disclosure agreement stating that I would never discuss what I had seen and done at Cambridge Analytica. Unbeknownst to me, the first trap in my future as a whistleblower had been set.

WHEN I MOVED BACK to Canada to work for Trudeau's research team, my days involved mostly conference calls and meetings, and for the most part I welcomed the stability and warmth of an environment that wasn't hostile, particularly one in which the boss wasn't hell-bent on psychologically abusing his staff.

In March 2016, a senior Canadian government official called me asking for a briefing that was slightly outside my mandate. He wanted a read on the Republican primaries in the United States, which were in full swing. Specifically, he wanted to know why Donald Trump was surging in the polls. On 1 March, Trump had taken the Republican primaries in seven out of eleven of the Super Tuesday states, and thousands of screaming supporters were packing his rallies across the country. The more outrageously Trump behaved, it seemed, the higher his poll numbers rose. At the 3 March presidential debate, he tangled with Senator Marco Rubio of Florida about, of all things, the size of his penis, with Trump boasting, 'I guarantee you there's no problem.' Two weeks later, Trump won four out of six states and territories in a single day – and Rubio dropped out of the race. Trudeau's people weren't worried – yet – but they were curious, because the reality TV star turned candidate struck them as ridiculous and weird. Why was he doing so well? What were the Americans thinking? Like many of their fellow Canadians, they took pleasure in smugly shaking their heads at their backward neighbours.

Canadians have a hard time understanding populism because, unlike America or Britain, it has never had any Rupert Murdoch-owned media. There is no Fox News or *The Sun* in Canada. Because of its more risk-conscious banking system, the country did not experience a housing crisis or financial crash. And unlike the rest of the OECD, Canada is the outlier where patriotism and support for immigration actually correlate positively with each other. So I would find myself repeating the same conversation over and over to baffled Canadians who simply could not understand how Brexit or Trump were even possible.

Pierre Trudeau, Canada's prime minister in the late 1960s and 1970s, once said that living so close to the United States was like 'sleeping with an elephant. No matter how friendly and even-tempered is the beast ... one is affected by every twitch and grunt.' Even if Trump didn't win – and few people at that point thought he would – his position on trade was already creating ripples. Trump hated the North American Free Trade Agreement (NAFTA) and was riling up the American electorate in states crucial to trade relations with Canada. The fear was not so much that Trump would win, but that the longer he lasted, the more his anti-NAFTA bombast might affect gubernatorial and legislative races in these states, which in turn would shift the national dialogue on trade.

The Cambridge Analytica saga was not yet lodged in the public consciousness, but it was no secret among my Canadian colleagues that my work for the firm had ultimately been passed on to certain US political campaigns. As Trump continued to gain ground, their curiosity grew. I described Cambridge Analytica's tactics of voter manipulation – how the firm identified and targeted people with neurotic or conspiratorial predispositions, then disseminated propaganda designed to deepen and accentuate those traits. I explained how, after obtaining people's data from Facebook, Cambridge Analytica could in some cases predict their behaviour better than their own spouses could, and how the firm was using that information to, in effect, radicalise people inside the Republican Party.

So while it was obvious that Trump touched a nerve with a certain percentage of American voters, Cambridge Analytica was working behind the curtain to raise his campaign to another level. A bunch of the people they were targeting were those who typically didn't vote Republican or didn't vote at all. They were trying to expand the electorate through this while at the same time they were committed to voter suppression. In particular, they focused on disengaging African Americans and other minority communities. One of the ways they did this was to peddle left-wing social justice rhetoric to depict Hillary Clinton as a propagator of white supremacy – while themselves working for a white supremacist. The aim was to move people from all demographics of a more left-wing ideology to vote for a third-party candidate, like Jill Stein.

I had started paying attention to candidate Trump when Cambridge Analytica sued me, because that was when I learned that the firm was working for him. At first, his campaign was a mess. But then he began repeating phrases like 'Build the wall' and 'Drain the swamp,' and he started rising in the polls. I called Gettleson and said, 'Well, this sounds eerily familiar, doesn't it?' – because these were the exact phrases CA had tested and included in reports sent to Bannon well before Trump announced. Meaning that, throughout the spring of 2016, when Cambridge Analytica was supposedly working for Ted Cruz, the fruits of its research seemed to be (*wink wink!*) making their way to Donald Trump.

As the primaries continued, it became apparent that Trump's chances of winning were increasing, and the attitude of people in Ottawa began to shift from 'He's crazy, ha ha,' to 'He's crazy … and he might become president of the elephant next to us.'

AS BREXIT LOOMED AND TRUMP gained ground, I realised it was time to speak up. I decided to connect with a couple of friends who worked in Silicon Valley. One of them – I'll call her 'Sheela' – knew someone at Andreessen Horowitz, the venture capital firm co-founded by the tech wunderkind Marc Andreessen. In the early nineties, Andreessen, along with Eric Bina, had created the Mosaic Web browser, forever changing the way people used the internet. Mosaic became Netscape, which became one of the earliest internet super-successes with its IPO in 1995. Since then, Andreessen had made hundreds of millions of dollars investing in companies like Skype, Twitter, Groupon, Zynga … and Facebook. He also sat on Facebook's board.

I flew to San Francisco in the spring of 2016 to start briefing relevant parties on what I'd seen at Cambridge Analytica. Sheela set up a meeting at the Andreessen Horowitz offices, on Sand Hill Road in Menlo Park. From the outside, the building looked like a slightly upscale suburban dentist's office, but inside, a rather bland lobby gave way to walls hung with fantastically expensive art. I met with Andreessen employees in a conference room and told them about Cambridge Analytica, the millions of Facebook profiles it had misappropriated, and the malicious way it was using the profiles to interfere with the election.

'Guys, you work for a major shareholder and board member,' I said. 'Facebook needs to be aware that this is going on.' They told me they would look into it. Whether that actually happened, I have no idea.

With a Facebook board member now apparently on the case, I went to a party in San Francisco's Mission District, where a Facebook

VP was expected to be among the guests. As it turned out, the party was filled with Facebook employees. The look was standard-issue Silicon Valley – form-fitted grey T-shirts – and it was hard to get through a conversation without hearing a progress report on a keto fasting diet, drinking Soylent meal replacements, and why food was becoming 'overrated'. Introduced as the guy from Cambridge Analytica, I quickly became the centre of attention, as they'd all heard so many rumours about the company. At the time, they all seemed to be aware of Cambridge Analytica, and I later found out that as far back as September 2015, Facebook employees discussed Cambridge Analytica internally and had asked for an investigation into the possible 'scraping' of data by the company. The employees reiterated their request for an investigation in December 2015, and would later be quoted in a Securities and Exchange Commission complaint filed against Facebook describing the firm as a 'sketchy (to say the least) data modelling company that has penetrated our market deeply.' But as I answered their questions, it was obvious that threats to democracy didn't interest them nearly as much as the mechanics of what Cambridge Analytica had pulled off. Even the Facebook vice president seemed mostly unfazed. If I had a problem with Cambridge Analytica, he said, then I should create a rival firm – respond to the Uber of propaganda by developing the Lyft. This suggestion struck me as perverse – not to mention irresponsible – coming from an executive at a company well positioned to take meaningful action. But that was how Silicon Valley operated, I soon realised. The reaction to any problem, even one as serious as a threat to the integrity of our elections, is not 'How can we fix it?' Rather, it's 'How can we monetise it?' They were incapable of seeing beyond the business opportunity. I had wasted my time. The US regulatory investigation that I later took part in eventually determined that at least thirty Facebook employees knew about Cambridge Analytica – but prior to the story being made public by my whistleblowing, the company did not put into place any procedures to report this to regulators.

Later, I was invited by Andreessen Horowitz staff to a private Facebook chat group called 'Futureworld', where executives from major Silicon Valley companies were discussing problems facing the tech sector, including issues I had raised. Andreessen also started talking to other executives in Silicon Valley about how their platforms were potentially being misused. He welcomed several other Silicon Valley notables to his house for a dinner group they began to refer to as *'the Junta'* – a reference to authoritarian groups that rule a country after taking power.

'It'll sure be ironic if the reason our correspondence lands on the government's radar,' one group member emailed Andreessen, was because 'their algorithms were triggered by our sarcastic use of the word "junta".'

IN EARLY SUMMER OF 2016, the Russia narrative started bubbling up. In mid-June, Guccifer 2.0 leaked documents that had been stolen from the Democratic National Committee. A week later, just three days before the Democratic National Convention, WikiLeaks published thousands of stolen emails, opening rifts between Bernie Sanders, Hillary Clinton and DNC chair Debbie Wasserman Schultz, who resigned almost immediately. And, of course, Nix eventually began asking around about Clinton's emails at the behest of Rebekah Mercer, eventually offering Cambridge Analytica's services to WikiLeaks to help disseminate the hacked material. I found out about this from a former colleague who was still with the firm and thought everything was getting out of hand.

As the Democrats tried to get their convention back on track, Donald Trump lobbed another metaphorical grenade at the party. At a news conference on 27 July, he casually invited Russia to continue its interference in the campaign. 'Russia, if you're listening,' he blustered, 'I hope you're able to find the thirty thousand emails that are missing' – a reference to emails that Clinton had deemed personal and deleted, rather than turning them over to investigators looking into her use of a private email server.

Over the summer and into the fall, Trump and Putin exchanged admiring comments, and I started thinking back on the weird Russia connections I'd noticed at Cambridge Analytica. Kogan's ties to St Petersburg. The meeting with Lukoil executives. Sam Patten's boasts about working with the Russian government. Cambridge Analytica's internal memos alluding to Russian intelligence. The Putin questions inexplicably inserted into our research. And even Brittany Kaiser's apparent connection to Julian Assange and WikiLeaks. At the time, I had thought these were strange events, each one incidental to the others. But now they began to seem like something more.

Trump became the nominee at the Republican National Convention on 19 July. If my hunch was correct, Cambridge Analytica was not only using the data tool I had worked on to manipulate American voters into supporting him, it may have been knowingly or unknowingly working with Russians to sway the election. Now that I was outside Cambridge Analytica looking in, it was as if I had X-ray vision. I knew the depths this company was willing to plumb, and the moral void at its core. I felt sick to my stomach. And I knew I had to tell someone – to raise the alarm.

I approached someone in the Trudeau government – I'll call him 'Alan' – and told him about my concerns. I started describing all the connections between Russia, WikiLeaks and Cambridge Analytica. I told him I had come to believe that Cambridge Analytica was a part of the Russia story and suggested that we share the details with someone in the US government.

We didn't want to cross a line, and we wanted to stay respectful of the US election. We were concerned that even though we only wanted to warn the United States about potential security threats, particularly from Russia, coming forward could be misinterpreted as a case of foreign actors attempting to interfere in the election – which we definitely were not. We settled on an alternative plan – a trip to Berkeley for a conference focused on data and democracy, where we could aim for a discreet chat with a couple of White House officials who we knew would be there.

The other person I had talked to extensively about all this was Ken Strasma, Obama's former targeting director. I had met with Strasma in New York and told him about Cambridge Analytica's data targeting. Since his firm had just provided microtargeting services for Bernie Sanders's 2016 campaign, naturally he was interested.

After Clinton clinched the nomination, in late July, Strasma called me and said, 'Now that we've lost, I'm going to see if I can talk with Hillary's data team.' He asked whether I'd be interested in meeting with them to outline my suspicions about what was happening with the Trump campaign. Yes, of course, I told him. Unfortunately, we were never able to connect with the Clinton team.

IN AUGUST, I TRAVELLED to Berkeley for a conference with some advisers from Justin Trudeau's office. We would be there for only a few days, so I asked another friend from Silicon Valley, whom I will call 'Kehlani', to help us arrange meetings. The most important one would be with the White House staffers.

I knew the meeting with the White House officials would be short, which meant we would have one shot to get our point across. And since it was likely that my audience would not be familiar with Cambridge Analytica, there was a good chance they wouldn't understand what we were talking about and would miss its significance. So I asked Kehlani to find a discreet setting where we could set up base and plan our meeting.

'How discreet do you want it?' she replied. 'I can get you a place that doesn't have cellphone reception.'

'A little unnecessary, but okay,' I said with a laugh. She gave us an address.

The following afternoon, we drove to where the GPS pointed us, which turned out to be the middle of a shipyard. Kehlani was waiting. She led us past a warehouse, down the dock. Things were getting weird, and they got even weirder when we had to step around giant harbour seals. Past them, we came upon a 135-foot Norwegian ferry, rusting in parts – it definitely wouldn't have passed inspection. The once-white ship had become grey, with barnacles stuck along its entire base. Someone threw down a ladder so we could climb up to the ship as it bobbed and swayed in the water.

Kehlani had found the most secure environment imaginable: a hacker ship. Anchored near San Francisco, the vessel was home to a handful of coders engaged in startups and other indeterminate tech activities. We didn't ask. In light of everything that was going on, it all felt perfectly apropos. We based ourselves on this ship throughout the trip.

Arriving at the conference the next day, we made arrangements for our unofficial meeting. Alan, in particular, was eager for us to emphasise that we were having this conversation in a personal capacity, not as representatives of Trudeau. The meeting would introduce employees of the Canadian government who were not representing that government to White House staffers who were not representing the White House. The topic was the US election and what was happening in the Republican Party with respect to Cambridge Analytica, including its massive surveillance database and its potential relationships with foreign intelligence agencies.

Someone from the White House group asked if we could talk outside, as they'd been cooped up all day at the conference. Which is how we ended up in a rather bizarre tableau: a group of high-level government advisers, crowded around a picnic table near the UC Berkeley campus, talking about Cambridge Analytica and Russian involvement in the US presidential election – all while weed-smoking, backpack-wearing students strolled past.

I got right to the point, warning the Americans about Cambridge Analytica's likely involvement in Russian interference. 'We know that there are individuals working on Trump's campaign who have ties to foreign intelligence services,' I said. 'They have built up a massive social media database which is being deployed on American voters.'

I detailed Cambridge Analytica's Russia connections and described the presentation Nix had made to Lukoil. I told them about the company's work to undermine people's confidence in the electoral process.

Their reply was ... ho-hum. One of the Americans said they couldn't do much, for fear of being accused of using the weight of the federal government to influence a vote. (I distinctly remember the phrase 'move the dial'.) The Obama group seemed extremely concerned about not tainting what they believed was a near-certain Clinton win. It seems ridiculous now, but at the time, the rumour was that, after his inevitable loss, Trump intended to launch Trump TV as a rival to Fox News. He was also expected to claim that the election had been rigged – that the deep state had influenced it, or Clinton had cheated, or both. Worried that Trump would use any irregularity to delegitimise the election, the Obama administration wanted to make sure it gave him nothing concrete to complain about.

When the White House guys told me about Trump TV, it made sense. I figured they knew what they were doing – and, after all, this was happening in their country, not mine. We all shook hands and went our separate ways. And this response was not unique. In early 2016, senior Facebook executives had identified that Russian hackers had probed the platform for individuals connected to the presidential campaigns, but chose not to warn the public or the authorities, as it would cause reputational issues for the company. (Facebook first publicly outlined the extent of the Russian information operations on its platform in September 2017, more than a year after it first identified the issue, and seven months after beginning to investigate what was called a 'five-alarm fire' of disinformation spreading on its site.) Ultimately, between the Democrats' indifference to the threat and Silicon Valley's inability to understand how to solve a problem without creating another 'Uber of *X*', my efforts to warn the Americans went nowhere. When you try to sound an alarm and people keep telling you 'Don't worry about it' or 'Don't rock the boat,' you start to wonder if maybe you're overreacting. I wasn't in the Clinton campaign or in the White House. I was just a guy in Canada, shouting into the wind.

The unfunny punch line, of course, is that, following the supreme reluctance of the Clinton and Obama teams to 'interfere' in the election, FBI director James Comey would sashay in and blow everything out of the water with his eleventh-hour decision to reopen the Clinton email investigation. At that point, back in Canada, I felt like I was watching a self-destructive friend finally tip over the edge. You can only stand there in horror, thinking, *I tried to tell you!* But in this case, the friend wasn't just burning down his own house. He was burning down the whole neighbourhood.

IN LATE AUGUST, Senator Harry Reid publicly urged the FBI to investigate Russian interference in the election. But at that time, most people still thought the race was Clinton's to lose. At the same time, Cambridge Analytica officially announced that it was working with the Trump campaign. This put nerves on edge in Ottawa, because I had been clear about the power and reach of Cambridge Analytica's data. The firm's work on behalf of Trump was worrying enough, but when you added the Russian connections into the equation, the whole situation seemed very alarming.

The possibility of a Trump win was the talk of Trudeau's office, though it was still mostly a laughing matter. On a scale of unimaginable to unthinkable to terrifying, just how unimaginable was it? We had one meeting where a couple of people were making fun of Trump. *God, these Americans! They outdo themselves every time!* And everybody laughed. Well, almost everyone. I wasn't laughing, because I understood the power of psychological warfare at scale.

The Germans have an expression called *Mauer im Kopf,* which translates roughly as *'the wall in the mind'*. After the East and the West reunified in 1990, the legal border between the two Germanys was dissolved. The checkpoints disappeared, the barbed wire was ripped away, and the Berlin Wall was finally taken down. But even fifteen years after reunification, many Germans still overestimated the distances between cities in the East and the West. There was, it seems, a lingering psychological distance that betrayed the nation's geography, creating a virtual border in people's minds. Although that wall of concrete and steel had long since crumbled, its shadow lived on, etched into the psyches of the German people. When this new candidate came out of nowhere, demanding that America build the wall, I knew that this was not a literal demand. Democrats and Republicans alike seemed at a loss for how to react to such an absurd campaign platform. But, unlike this new dark horse, they could not peer into what was happening in the minds of America. They could not see that those people were not merely demanding a physical wall. It was not about building a literal wall – the very idea of the wall itself was enough to begin to achieve Bannon's aims. They were demanding the creation of America's very own *Mauer im Kopf.*

Alan wasn't laughing, either. In one meeting he said, 'I actually think Trump can win.' People looked at him and rolled their eyes, and someone said, 'Come on'. Then he looked at me, and I said, 'Yeah, I think he can win, too.' That was the moment when I truly grasped, to my utter horror, that the tools I had helped build might play a pivotal role in making Donald Trump the next president of the United States.

A couple of weeks later, a letter from Facebook arrived at my parents' house. How Facebook got their address, I had no idea. My mother forwarded the letter to me. It had been sent by a law firm Facebook had hired, Perkins Coie, which was the same firm that the Clinton campaign was using to fund the private investigation into what later became the Trump-Russia dossier. Facebook's lawyers sought to confirm that the data Cambridge Analytica had obtained was used only for academic purposes, and that it had been deleted. Now that Cambridge Analytica was officially working on the Trump campaign, Facebook had apparently decided it would be bad optics to have millions of its users' personal profiles raided for political gain, not to mention Cambridge Analytica's astonishing commercial enrichment. The letter made no mention of using company data to turn the world inside out. And, of course, the letter was preposterous, a feeble gesture, because Facebook had given the data-harvesting application used by Cambridge Analytica *express permission* to use the data for non-academic purposes – a request I had made specifically to the company while working with Kogan. I was even more confused at Facebook's feigned pearl-clutching reaction, as in or around November 2015 Facebook hired Kogan's business partner Joseph Chancellor to work as a 'quantitative researcher'. According to Kogan, Facebook's decision to hire Chancellor came after the company was told about the personality profiling project. Later, when the story became public, Facebook played the role of shocked victim. But what it did not make clear was that it was content to hire someone who had actually worked with Kogan. However, in a statement Facebook later said, 'The work that he did previously has no bearing on the work that he does at Facebook.'

There was no way Cambridge Analytica had deleted the Facebook data, of course. But I had left the firm more than a year earlier, and had been sued by them, and wanted absolutely no part of speaking for them. I replied, saying that I no longer had the data in question, but I had no idea where the data was, who else got ahold of it, or what Cambridge Analytica was doing with it – and neither did Facebook. But I was paranoid enough about being connected to Cambridge Analytica in any way that, instead of just popping the letter into the outgoing mail from the Canadian Parliament, I walked downtown to mail it. I didn't want the taint of Cambridge Analytica anywhere near my work for Trudeau.

On 22 September, 2016, Senator Dianne Feinstein and Congressman Adam Schiff released a statement saying that Russia was attempting to undermine the election. And in the first presidential debate, on 26 September, Hillary Clinton sounded the alarm. 'I know Donald is very praiseworthy of Vladimir Putin,' she said. Putin had 'let loose cyber attackers to hack into government files, to hack into personal files, hack into the Democratic National Committee. And we recently have learned that this is one of their preferred methods of trying to wreak havoc and collect information.'

'I don't think anybody knows it was Russia that broke into the DNC,' Trump responded. 'She's saying Russia, Russia, Russia. Maybe it was. I mean, it could be Russia, but it could also be China, could also be lots of other people. It also could be someone sitting on their bed that weighs four hundred pounds, okay?'

On 7 October, less than an hour after the *Access Hollywood* tape of Trump saying 'grab 'em by the pussy' became public, WikiLeaks began publishing emails hacked from the account of Clinton campaign chair John Podesta. They would continue to release emails, bit by bit, until Election Day, and the fallout for the Democrats was disastrous. Scandals erupted over details of Clinton's speeches to Wall Street, among other revelations, and the lunatic fringe of the alt-right used the emails to fuel the deranged theory that the Clinton campaign, at the highest levels, was involved in a child sex ring being run out of a Washington, DC, pizza parlour. My mind kept returning to the connections among Cambridge Analytica, the Russian government and Assange. Cambridge Analytica seemed to have its dirty hands in every dirty part of this campaign.

ON THE NIGHT OF the election, I was at a watch party in Vancouver. We had CNN on one giant screen, and other news channels on smaller screens, and at the same time I was on the phone with Alistair Carmichael, the MP from the Shetlands, whom I'd grown close to during my time in London. I was getting the US, Canadian and British reactions in real time as the numbers started looking more and more dire for Clinton. The moment CNN projected Trump as the winner, a shocked pall fell over the room.

My phone started buzzing with text messages from people who knew about my work with Cambridge Analytica. Some bewildered supporters at Hillary Clinton's victory party began directing their anger toward me. I don't recall many of the specifics, just an overwhelming tone of rage and despair. There is one comment I do remember, though, because it cut me to the bone. A Democrat I was friendly with wrote, 'This may have been just a game to you. But we are the ones who have to live with it.'

That night and the day after, Trudeau's advisers were in meltdown, because everything they thought they knew about the elephant to the south had suddenly changed. Was Trump going to cancel NAFTA? Would there be riots? Was Trump a plant for the Russians – an actual Manchurian candidate? People were desperate for answers, and because I was the only guy who knew anything about Steve Bannon, who was now in a position of immense power, they kept asking me what was going to happen. They wanted advice on how to deal with these new alt-right advisers whom they would soon be negotiating with on issues of massive national (and international) importance. All I could think was: *Holy fuck!* The man I'd met in a hotel room in Cambridge three years earlier now had the ear of the future president of the United States.

When Carmichael called me the day after the election, it was a relief to hear his calming Scottish brogue. 'Let's think carefully about what you're going to do,' he said. Over the years, I had confided everything about Cambridge Analytica to Carmichael, who was one of the few people in the world I trusted completely. He knew me well enough to understand that I was unlikely to be able to sit by while Trump and Bannon took control after a tainted election. The stakes had been raised exponentially. The ridiculous reality TV star was no longer just a shameless rabble-rouser. He was going to be the leader of the free world.

Throughout November and December, I contemplated what I might say, and to whom I might say it. Trump's election still didn't seem real, because Obama remained president. It was as if the whole world was holding its breath, waiting to see what would happen after 20 January.

Before the election, friends in the Democratic Party had offered to help get me tickets to Clinton's inaugural ball. But instead of flying to DC to party with deliriously happy Democrats, I watched the sparsely attended Trump inauguration on CNN. And that was when I saw something that was almost hard to believe. There they were: Bannon, looking like a windswept gremlin. Kellyanne Conway, whom I'd met through the Mercers, in full Revolutionary War cosplay. Rebekah Mercer, wearing a fur-lined overcoat and Hollywood-starlet sunglasses. And then I remembered what Nix had said to me a couple of years earlier, at the restaurant where I told him I was quitting Cambridge Analytica. 'You're only going to understand it when we're all sitting in the White House,' he'd said. 'Every one of us, except for you.' Well, Nix wasn't in Washington, but the rest of them certainly were.

That January, Bannon was appointed to the National Security Council. Now Carmichael's warning for me to 'be careful' seemed even more apt, as Bannon had the levers of the entire American intelligence and security apparatus at his disposal. If I pissed Bannon off, either by whistleblowing or through some other provocation, he had the capacity to destroy my life.

Just as worrisome, Bannon was in a position to help arrange for Cambridge Analytica to get contracts with the US government. Cambridge Analytica's parent company, SCL Group, was already working on projects for the US State Department. This meant that CA could access the US government's data, and vice versa. To my horror, I realised that Bannon could be creating his own private intelligence apparatus. And he was doing it for an administration that didn't trust the CIA, the FBI, or the NSA. It felt like I was living in a nightmare. Worse, it felt like Richard Nixon's wet dream. Just imagine if Nixon had had access to that kind of intimate, granular data on every single American citizen. He wouldn't have just *fucked rats,* he would have fucked the *whole Constitution*.

Government entities normally need a warrant to collect people's private data. But because Cambridge Analytica was a private company, it wasn't subject to that check on its power. I started to recall the meetings with Palantir employees, and why some of them were so excited by Cambridge Analytica. There was no privacy law in the United States to stop Cambridge Analytica from collecting as much Facebook data as it could. I realised that a privatised intelligence unit would allow Bannon to bypass the limited protections Americans had from federal intelligence agencies. It occurred to me that the deep state wasn't just another alt-right narrative; it was Bannon's self-fulfilling prophecy. He wanted to *become* the deep state.

11

COMING OUT

TWO MONTHS AFTER TRUMP'S INAUGURATION, ON THE MORNING of 28 March, 2017, I woke up slightly groggy from another late night working on a briefing. It was just after six in the morning. As I stood in my underwear, waiting for the coffee to brew, I opened up Facebook and saw a private message from an account called 'Claire Morrison'. I clicked on the profile, and there was no profile photo.

> Hi Christopher, I hope you don't mind me getting in touch. I'm actually a journalist ... called Carole Cadwalladr. I have been trying to talk to previous employees of Cambridge Analytica/SCL to try and build up a more accurate picture of how the company works etc and you've been described to me as the brains of the operation . . .

This has to be Cambridge Analytica, I thought. Not again. I just *knew* it had to be Cambridge Analytica messing with my head. No journalist had ever reached out to me, all of my warnings had been ignored, and this was exactly the kind of shit Nix would pull. I wanted no part of this faceless 'Claire Morrison', absent solid proof of who was behind the profile. So I replied that I needed proof that she was actually a *Guardian* journalist.

That same day, Cadwalladr sent me a long message from her *Guardian* email address about Vote Leave, BeLeave, Darren Grimes,

Mark Gettleson, and how everything in those campaigns seemed to be routed through this little company in Canada, AggregateIQ. She had been told that I knew the people and the companies involved. Cadwalladr wrote that she had been investigating AIQ and Brexit and that, one evening in early 2017, a source had tipped her off with a bizarre clue. The telephone number listed on the official expense returns for AIQ appeared on an archived version of SCL's website as the phone number for 'SCL Canada'. At the time, there was almost no public information on SCL, save for a 2005 *Slate* article Cadwalladr had found about the firm, titled 'You Can't Handle the Truth: Psy-ops Propaganda Goes Mainstream'. The article began with a scenario where a 'shadowy media firm steps in to help orchestrate a sophisticated campaign of mass deception'.

As Cadwalladr kept pulling the threads of this increasingly bizarre story, she found a former SCL employee in London who was willing to talk. The source was insistent that they meet somewhere discreet and that she had to keep everything off the record. The source was afraid of what the firm would do if they found out the two were talking. Cadwalladr listened as the source told her some wild stories about what SCL had done in Africa, Asia and the Caribbean – honey traps, bribes, espionage, hackers, strange deaths in hotel rooms. The source told her to find someone by the name of Christopher Wylie, as he was the one who had recruited AIQ into the Cambridge Analytica universe. In examining all the complex relationships between all of these people and entities – Vote Leave, AIQ, Cambridge Analytica, Steve Bannon, the Mercers, Russia and the Trump campaign – Cadwalladr saw that I was right in the middle of all of them. Seeming to pop up everywhere, I looked like the Zelig of 2016 to her.

At first, I didn't want to talk to Cadwalladr. I had no interest in being at the centre of some massive *Guardian* exposé. I was exhausted, I had been burned over and over, and I wished that I could just put the Cambridge Analytica ordeal in the past. On top of that, Cambridge Analytica was no longer just a company. My old boss Steve Bannon was now sitting in the White House and on the National Security Council of the most powerful nation on earth. I had seen what happened to whistleblowers like Edward Snowden and Chelsea Manning when they were at the mercy of the full force of the US government. It was too late to change the outcome of the Brexit vote or the US presidential election. I had tried to warn people, and no one seemed to care. Why would they care now?

But Cadwalladr cared. When I read what she had already published, I could see that she was on the trail of Cambridge Analytica and AIQ but had not yet cracked how deep their misdeeds actually went. After hesitating for a couple of days, I emailed back and agreed to talk, but strictly off the record. When it was time for our call, my heart was racing. I fully expected an unpleasant conversation in which she would make accusations and listen half-heartedly to my responses, after which she'd write whatever she wanted.

Instead I got a woman saying, 'Oh, is that Chris? Oh, hi!' I heard barking in the background, and she said, 'Sorry, I've just taken my dog for a walk, and I'm making some tea.' I started to say something, but I heard her making cooing sounds to her dog. I had intended to give Carole about twenty minutes, but four hours later we were still on the phone. It must have been well past midnight in London, but the conversation just kept going and going. This was the first time I had really talked to anyone about the totality of what had happened. What, she asked me, is Cambridge Analytica?

'It's Steve Bannon's psychological mindfuck tool,' I told her bluntly.

Even a well-informed journalist like Cadwalladr struggled at first to understand all the layers and connections of the Cambridge Analytica narrative. Was SCL part of Cambridge Analytica, or the other way around? Where did AIQ fit in? And even when she had the basic details nailed down, there was still so much more to share. I told her about psychometric profiling, information warfare and artificial intelligence. I explained Bannon's role and how we had used Cambridge Analytica to build psychological warfare tools to fight his culture war. I told her about Ghana, Trinidad, Kenya and Nigeria and the experiments that shaped Cambridge Analytica's data targeting arsenal. Finally she began to understand the scope of the company's wickedness.

With the title 'The Great British Brexit Robbery: How Our Democracy Was Hijacked,' that first article came out 7 May, 2017. It caused a sensation, becoming the most read piece on the *Guardian*'s website that year. Cadwalladr's reporting was solid, but she had only just started to skim the surface of a much murkier story. On 17 May, Robert Mueller was appointed special counsel to oversee the investigation into Russia and the Trump campaign. It was beginning to become apparent that there was an appetite among Democrats and even some Republicans to get to the bottom of why Trump had so drastically fired FBI director James Comey after telling him to drop the investigation into his former national security adviser, Michael T. Flynn – who, it turned out, had a consulting agreement with Cambridge Analytica. The full story went far beyond Brexit – it was about Bannon, Trump, Russia and Silicon Valley. It was about who controls your identity, and the corporations that traffic in your data.

But I had a problem. If I was to help get this information out and persuade others at Cambridge Analytica to come forward, I couldn't do it in Canada. I confided in some of Justin Trudeau's team. They immediately understood the gravity of the situation and encouraged me to come forward and go to the UK to work with the *Guardian*. So I did.

I HAD NO PLAN and didn't even have a flat sorted yet, so I flew to Alistair Carmichael's constituency in the Shetland Islands, the most northerly of the British Isles, once part of ancient Norse kingdoms before its annexation by Scotland. At the airport, when I stepped from a tiny propeller plane with my life packed into a single bag, Carmichael was waiting, ready to take me to a guesthouse – but not without a scenic detour. As the local member of Parliament and an extremely proud Scotsman, he was keen to first give me a tour of the island, rain and harsh wind be damned. We were surrounded by sheer cliffs, Shetland ponies, and sheep roaming nearby when he asked what I planned to do.

'I don't have a plan yet,' I answered. 'I'm seeing Carole next week … Do you think this is a good idea, Alistair?'

'No – it's crazy!' he shot back, almost shouting. Then he fell silent. 'But it's important, Chris. All I can say is that I will do what I can to help.' There are few politicians I would walk through miles of cold, wet grass in the freezing north of Scotland for. But Alistair had always been someone I could rely on. He'd become a confidant, a mentor and a friend.

A couple of weeks later, Cadwalladr and I finally met in person, near Oxford Circus in London, at the Riding House Café – a large modern space with crimson sofas by the windows and a bar with bright turquoise stools. Cadwalladr was waiting for me inside, looking like a biker chick with her tousled blond hair, sunglasses, leopard print top and well-worn leather bomber jacket. From across the street, I could see her in the large window of the restaurant. Unsure that this woman was actually the same *Guardian* journalist I had been speaking with for months on the phone, I searched for more photos of Cadwalladr on my phone, and held up the photos to compare with the woman sitting inside. When she saw me, she jumped up and exclaimed, 'Oh my God! It's really you! You're taller than I imagined!' She got up and gave me a hug. She told me that the *Guardian* wanted its next story to be about how Cambridge Analytica had collected the Facebook data and asked if I would be willing to go on the record.

It would not be an easy decision. If I went public, I risked the wrath of the president of the United States, his alt-right whisperer Steve Bannon, Downing Street, militant Brexiteers and the sociopathic Alexander Nix. And if I told the whole truth behind Cambridge Analytica, I risked angering Russians, hackers, WikiLeaks and a host of others who'd shown no compunction about breaking laws in Africa, the Caribbean, Europe and beyond. I had seen people face serious threats to their safety; several of my former colleagues had warned me to be extremely careful after I left. Before joining SCL, my predecessor, Dan Mureşan, had ended up dead in his hotel room in Kenya. This was a decision that I could not take lightly.

I told Cadwalladr I would think about it and continued to give her information. But then any trust I had in the *Guardian* was wrecked when the paper failed to stand by its own reporting. Cadwalladr had opened her 7 May article with the story of how Sophie Schmidt – daughter of Google CEO Eric Schmidt – had introduced Nix to Palantir, setting off the chain of events that led to SCL's foray into data warfare. I knew the story, though I wasn't the source for it; someone else had told Carole about it. The article was true. In fact, I had emails about Sophie Schmidt's involvement in SCL. The story wasn't remotely libellous, but Schmidt threw a battalion of lawyers at the *Guardian*, with the threat of a time-consuming and expansive legal battle. Instead of fighting an obviously spurious lawsuit, the paper agreed to remove Schmidt's name several weeks after publication.

Then Cambridge Analytica threatened to sue over the same article. And even though the *Guardian* had documents, emails and files that confirmed everything I had told them, they backed down again. Editors agreed to flag certain paragraphs as 'disputed', to appease Cambridge Analytica and mitigate the paper's liability. They took Cadwalladr's well-sourced story and watered it down.

At this point, my heart sank. I thought, *All right, I've just moved back to London, I haven't got a job, and I'm being asked to put my neck on the line for a newspaper that won't even defend its own journalism.* An additional complication was the super NDA that prohibited me from revealing details about my work at Cambridge Analytica. The whole point of Cambridge Analytica making me sign it was that it seriously increased my legal liability, and I had no doubt that my old employers would attempt to sue me into oblivion were I to breach the agreement. My lawyers said I had a strong defence – that by giving the *Guardian* this information, I was exposing unlawful behaviour. But a good defence does not prevent a lawsuit from being filed in the first place, and fighting Cambridge Analytica in court would mean hundreds of thousands of pounds in legal fees – money I didn't have.

All the same, I remained determined that the full story come out. My best course of action, I soon discovered, ran through Donald Trump's hometown. Carole passed me on to Gavin Millar QC, a well-known barrister in London at Matrix Chambers who had worked on the Edward Snowden case for the *Guardian*, and he suggested that I give the story to an American newspaper. The First Amendment provided US newspapers with much stronger defences against accusations of libel, he said. *The New York Times* was far less likely to back down than the *Guardian* had been, and it would never delete parts of articles after the fact. This was a brilliant suggestion. It also ensured that the story would get as much play in the United States as in Britain.

I told the *Guardian* that I planned to share the story with *The New York Times*. They were not pleased, arguing that if we waited any longer, interest in the story would dry up, or someone else would publish before us. But the choice was mine, not theirs, and I stood my ground. I would give reporters at both papers the same information, with the condition that they publish on the same day – and only after I gave the go-ahead. There was too much at stake, and the *Guardian*'s actions in the Schmidt case had made me wary of the risk of Britain's extremely plaintiff-friendly libel laws. I reiterated to the paper's editors that I would not be cooperating or handing over documents until there was an agreement with *The New York Times*. Cadwalladr was actually completely supportive of letting the *Times* in, and the *Guardian,* not having much choice in the matter, acquiesced. The editors cringed at the idea of sharing with their rivals, but, to their credit, they swallowed their pride and set up a meeting with *Times* editors in Manhattan to start the discussions about how this was all going to play out. The papers reached a tentative agreement in September 2017, and shortly thereafter I met the reporter *The New York Times* had assigned to the story.

ON THE DAY I was scheduled to meet the American journalist, I walked into the buzzing lobby bar of the Hoxton Hotel, in Shoreditch, and spotted Cadwalladr, who waved me over to her table. She was sitting across from Matt Rosenberg of the *Times*. Completely bald, slightly beefy, and apparently divorced, he was quite fetching.

'So you're the guy?' Rosenberg said as he rose from his seat to shake my hand. 'I'm assuming,' he said, 'we need phones away?'

We all pulled out our Faraday cases, which prevent phones from receiving or transmitting electronic signals. All my meetings with journalists started with this ritual. We then zipped the cases into a soundproof bag I had brought in case there was preinstalled listening malware that could turn itself on without remote activation. With my former Cambridge Analytica associates now working in the Trump administration, and given Cambridge Analytica's history with hackers and WikiLeaks, we had to be extremely careful.

After more than two hours of conversation about my experience with Cambridge Analytica, Rosenberg said that he had enough to go back to his editors. He ordered some wine for the table and then shared war stories from his time in Afghanistan. He seemed like a straight-up decent guy, and I felt hopeful that maybe this was going to work out. Before the meeting broke up, he gave me his card: 'Matthew Rosenberg. National Security Correspondent. *New York Times*.' He had scribbled a number on the back. 'That's for my burner phone. Call me on Signal. It'll be good for a few weeks.'

With *The New York Times* now involved, I began connecting journalists with other former Cambridge Analytica employees, and the journalists flagged a recurring theme: everyone thought that if they could talk directly to Nix, he wouldn't be able to help himself – he'd start bragging about Cambridge Analytica's operations to further wank his already overinflated ego. Although this was undoubtedly true, it seemed like a bad idea to tip him off about planned exposés.

'Maybe I should try to interview him,' Cadwalladr said to me one afternoon. Then she came up with a better idea: catching him in the act. If we put Nix in a situation where he was trying to win potential clients, he was sure to blab about his shady tactics in the hope of impressing them. I'd seen him do this at least a dozen times. And if we managed to get him on tape doing it, that would prove to the world that my accusations were true. So now, in addition to the *Guardian* and *The New York Times,* we decided to approach Channel 4 News. As a public television service, Channel 4 had a statutory mandate to represent more diverse, innovative and independent programming than the BBC, which tended to be extremely risk averse in breaking new stories.

One afternoon in late September, Cadwalladr and I met Channel 4's investigations editor, Job Rabkin, and his team at the back of an empty pub in Clerkenwell a few blocks from their studios. Cadwalladr introduced us, and Rabkin began describing his team's experience with undercover work. When I started telling him about Cambridge Analytica's projects in Africa, Rabkin's eyes widened. He interjected, 'This sounds so twisted and *colonial*.' Rabkin was the first journalist to use that word with me – 'colonial'. Most of the people I told about Cambridge Analytica were fascinated with Trump, Brexit or Facebook, but whenever I got onto the topic of Africa, I usually was met with shrugs. *Shit happens. It's Africa, after all*. But Rabkin got it. What Cambridge Analytica was doing in Kenya, Ghana, Nigeria – it was a new era of colonialism, in which powerful Europeans exploited

Africans for their resources. And although minerals and oil were still very much part of the equation, there was a new resource being extracted: *data*.

Rabkin pledged the full support of the Channel 4 investigations unit, adding that his team was willing to take the risk of going undercover to get inside Cambridge Analytica. I began working with them on the operation, which I felt sure would truly blow the lid off of Nix's corrupt tactics. But this would be an incredibly complicated and delicate undertaking, with disastrous results if Nix somehow discovered what we were up to.

With so many moving parts, whistleblowing was turning into a full-time job. I was also setting myself up for a storm of legal trouble if anything was done improperly, so I texted the lawyers who had helped me over the summer. Their answer was the last thing I wanted to hear. This was too much for pro bono support; I needed to either find cash or new lawyers. I was devastated. I was out of work, in the thick of an extremely complicated mess, and risking serious legal jeopardy – all without a lawyer. But as with so many things in life, sometimes you get lucky and bad news leads you somewhere amazing. And this was one of those moments, because it led me to Tamsin Allen.

Gavin Millar heard about what had happened and referred me to Allen, a top UK media lawyer at the firm Bindmans LLP who was an expert in defamation and privacy, in the fall of 2017. Her list of clients included ex-MI5 spies and celebrities who'd had their phones hacked in the infamous News Corp case. She seemed like a perfect fit for my dilemma, and when I met her, we immediately clicked. Growing up, Allen was once expelled from school for skinny-dipping, and just as the 1980s punk scene was taking off, she moved to London, where she lived with squatters in Hackney. 'I have so many stories that are completely untellable,' she recollected one late evening as we were prepping evidence. Allen was a rebel herself, and she was unfazed by a pink-haired guy with a nose ring recounting bizarre stories of spying, hackers and data manipulation. In my journey to become a whistleblower, Allen became my number one ally.

Allen recognised that my interests weren't entirely aligned with those of Channel 4, the *Guardian* and *The New York Times*. The reporters were focused on the scoop of the year, maybe the decade, while I needed to tell this vitally important story and simultaneously steer clear of legal jeopardy. She counselled that I stay razor-focused on the public interest aspect of the Cambridge Analytica story, particularly because of the super NDA, as British law allows a defence to breach of confidence if it was required to reveal illegality or was manifestly in the public interest. We had a lot of discussions about what 'the public interest' was and how we could hew to that line, avoiding revealing anything too gossipy or that would risk legitimate national security interests of the British or American governments, but Allen warned that even if we kept to an airtight legal standard, Cambridge Analytica would still probably sue me. She told me that Facebook also might file suit, and their resources were nearly limitless. And she said it was possible that Facebook or CA might apply for an injunction to prevent publication. Such injunctions are almost unheard of in the United States but are not uncommon in the UK. Fighting the injunction would be time-consuming, and even if we ultimately won, the British journalists might get cold feet and pull out – Allen told me she had seen it happen many times.

But those were simply the legal scenarios. The story involved many characters who had a history of operating outside of the law, and Allen was becoming concerned for my personal safety. At one of our early meetings, she asked whether I had family in London and what safety precautions I'd taken. 'Who are you going to ring in an emergency?' she asked me. We had to create a plan. But as time went on and we became increasingly devoted to each other, I decided that Allen would be the one I would call should things go sideways.

With my legal situation sorted, I started having conversations with Sanni about what had transpired between BeLeave and Vote Leave. He was very forthcoming. Without quite grasping the deeper implications of what he was describing – collusion and cheating – he outlined an arrangement in which Vote Leave wired hundreds of thousands of pounds through BeLeave to an AIQ account. After I helped him see the underlying offenses, Sanni finally understood that he had been used. He had no idea that AIQ was part of Cambridge Analytica, and he was visibly disgusted when I told him about the videos that AIQ had handled for CA during the Nigerian election.

A few days later, he showed me a shared drive containing strategy documents for BeLeave, Vote Leave and AIQ. Under British law, this was evidence of unlawful coordination. In the activity log, someone had been using the administrator's account to delete the names of leading Vote Leave officials from the drive. The deletions were made the same week the Electoral Commission launched its investigation of the campaign, Sanni told me. Vote Leave has since claimed this was a data clean-up, but it looked to me as if Vote Leave was trying to delete evidence of the over-spending and had potentially committed yet another crime: destruction of evidence. It began to look like an attempted cover-up, and when he showed me who else was on this drive, it became even more serious. Two of the accounts on the shared drive were those of two senior advisers who were now sitting inside the office of the prime minister, advising on the Brexit negotiation process. I emphasised to Sanni that he might be in possession of evidence of a crime – several crimes, in fact – and that he needed to be extremely careful lest he end up in serious trouble. He was already aware of my cooperation with the *Guardian* and *The New York Times*. As he grasped the enormity of what he discovered, Sanni agreed to meet with Cadwalladr, to tell her what he knew. And I also connected him to Tamsin Allen for independent legal advice.

In the beginning, Allen worked on a pro bono basis. But as the complexity of the situation grew, she could no longer give me the hours I needed without some sort of payment. She was also concerned about what would happen if Cambridge Analytica took me to court after they found out that I was going public. Allen refused to drop me over money, but we had to get creative. We decided to approach some of her well-connected contacts, as Allen knew it would be important to create a body of support. The first was Hugh Grant – yes, *the* Hugh Grant, the star of *Four Weddings and a Funeral* and *Bridget Jones's Diary*. Over dinner, Allen explained my predicament. Sanni joined us to go through what had happened on Vote Leave. Grant was warm and caring and seemed just like many of the characters he's played. Grant had also had personal experience with stolen data – the Rupert Murdoch-owned *News of the World* had hacked his phone messages. He was taken aback by the scale of Cambridge Analytica's activities and said he would help think of who could assist us.

The crucial piece of support came a few weeks after we were introduced to Lord Strasburger, a Lib Dem who sat in the House of Lords and is the founder of Big Brother Watch, a privacy activist group. He, in turn, connected me to an exceptionally wealthy individual who came to London to meet me. I asked him why he wanted to help, and he told me that it was because he knew the history of Europe. He said he knew what happens when everyone is catalogued. Privacy is essential to protect us from the rising threat of fascism, and so he said he would help me. A few days later he pledged funds to help me and gave me the backing I needed.

This was just part of the assistance that allowed me to get through the whistleblowing ordeal in one piece. As I prepared to speak out against political and corporate Goliaths, I was a David now backed by committed lawyers and journalists, a legal defence fund, and a huge amount of moral support. So often, whistleblowers are framed as lone activists standing up to giants for what's right. But in my case, I was never alone, and I got incredibly lucky on several occasions. Without this help, I never would have been able to come forward.

IN OCTOBER 2017, Allen and I met with the Channel 4 News producers, Job Rabkin, and his editor, Ben de Pear. I described Nix and told them about the kinds of illegal activities he regularly engaged in. They were intrigued by the notion of catching him on tape, but as we discussed the details of how such a sting might work, they wondered whether it would be too complicated to succeed. Permission would have to come from the top legal team at the channel, and they could easily deem the undertaking too risky, both legally and reputationally, if it backfired.

We began working with their lawyers on an extensive legal preparatory document. Britain has laws protecting these types of sting operations, but journalists must show that what they're proposing is in the public interest, that they're not going to entrap anyone, and that the sting is going to reveal probable crimes. Preparing this document in advance would protect Channel 4 in the event Nix sued.

The sting would require microscopic attention to detail. I called Mark Gettleson, who without hesitation agreed to help. Nix would have to believe that the people he was meeting with were clients, that the project he was being asked to undertake was real, and that the conversation he was having was completely private. The person playing the 'client' would have to be extensively briefed on how Nix operated. He would have to know exactly what to ask for and would also have to be well versed in the political situation of whichever country we chose to have the 'project' based in.

We decided to set the scenario in Sri Lanka, for a couple of reasons. One: SCL had done work in India and had an office there, so a neighbouring country would feel familiar enough to Nix. And two: the labyrinthine nature of Sri Lankan politics and history made it easier to create a fake political scenario loosely based in reality. Whatever project we created would have to involve enough real players that when a Cambridge Analytica assistant did a bit of Googling before the meeting, it would still seem legit and eventually pass their due diligence procedures.

After Channel 4 hired a Sri Lankan investigator to play the client, 'Ranjan', Gettleson and I coached the Channel 4 team on Nix's habits and peculiarities, walked them through how Cambridge Analytica vetted prospective clients, and showed them emails from Nix to help them ascertain how he and the company operated. There would be four meetings altogether – three preliminary ones with other Cambridge Analytica executives and then a final one to close the deal with Nix. Ranjan would have to let Nix come up with illegal ideas on his own so there could be no possible suggestion of entrapment.

Ranjan was to play an agent representing an ambitious young Sri Lankan who'd travelled to the West, made a lot of money, and now wanted to return home to run for political office. But because of a family rivalry, a particular minister in the government had frozen his family's assets. He would use the name of an actual minister and throw in enough factual detail about Sri Lankan politics that Nix and the other executives would buy into the whole scenario. Channel 4 had to do a huge amount of detailed advance research, because any misstep could potentially blow the whole sting up. The carrot for Cambridge Analytica was 5 per cent of the value of the man's assets, if they succeeded in getting the (imaginary) funds released. We knew Alexander wouldn't be able to resist.

At the first two meetings, Ranjan met with chief data officer Alexander Tayler and managing director Mark Turnbull in private rooms at a hotel near Westminster. The executives pitched Cambridge Analytica's data analysis work and suggested intelligence-gathering services, but nothing concrete came out of the meetings. They seemed cagey, hedging in how they talked about what Cambridge Analytica really did. Channel 4 was frustrated, but we had an idea for how to fix it.

We realised that whenever guys like this go into a private hotel room, they assume that it's bugged, so Channel 4 had to figure out how to have these meetings in a public place. The Channel 4 executives pushed back, saying the logistics would be impossible. If we tried to record a meeting in a restaurant or bar, the noise might drown out the audio. Also, where would we put the cameras, to ensure we'd get video recordings of the executives? We couldn't just lead them to a specific table – that would be too suspicious.

The Channel 4 team, to their credit, made an audacious decision. They rented out a large part of a restaurant, filled it with people hired to have lunch and talk quietly, and aimed dozens of hidden cameras at all the available tables. Nix and the other executives could choose whichever table they liked, which would help make them feel less guarded. However, pretty much everything around them was a camera – even some of the table settings, handbags, and 'guests' sitting around them would be recording the conversation.

Two meetings took place at this restaurant. At the first, Turnbull laid the groundwork for some of the more questionable services Cambridge Analytica offered. He told Ranjan that Cambridge Analytica could do some digging about the Sri Lankan minister, saying they would 'find all the skeletons in his closet, quietly, discreetly, and give you a report.' But he pulled back toward the end, saying that 'we wouldn't send a pretty girl to seduce a politician, and then film them in their bedroom and then release the film. There are companies that do this, but to me, that crosses a line.' Of course, just by describing what Cambridge Analytica supposedly wouldn't do, he succeeded in putting the idea on Ranjan's plate.

Finally, weeks after the sting began, Nix made his entrance. For this fourth meeting, Channel 4 took extra care to make sure that everything was arranged flawlessly. All the tables were bugged, and cameras were set up around the room. There were hidden cameras in the handbags of a couple of women having lunch at the next table. Everything was ready, and we held our breath that Alexander wouldn't cancel or postpone.

He didn't. He dug his own grave. Ranjan did a perfect job of asking the right questions and showing interest at the most opportune moments. And Alexander, bless him, just walked right in and opened his big mouth.

TWO MONTHS WOULD PASS before Channel 4 was ready to show us the tape from the restaurant. One morning in early November, I had an appointment with Allen, and I decided to walk because it was a sunny and crisp autumn morning. As I waited for her in the reception lobby, I noticed that I had received several new messages from a strange number. I opened them and reflexively yelled, 'What the fuck!' The receptionist stood up and asked if I was okay, and I said no. The messages were photos of me walking that morning. Someone had followed me to my lawyer's office, and they wanted me to know.

We suspected that Cambridge Analytica may have found out that I had moved back to London and had hired a firm to find out what I was doing. From that point on, Allen said we needed to change my routines – where I went and how I met with lawyers. A few days later, Leave.EU posted on Twitter a video clip from the movie *Airplane!* depicting a 'hysterical' woman being repeatedly beaten, with Cadwalladr's face superimposed. In the background played the Russian national anthem. She told me she had found out that they might be using a private intelligence firm to investigate her and warned me that if they had been following her around, they might have seen me too and connected the dots. If Cambridge Analytica found out what I was doing, Allen cautioned me, the company could go to court and apply for an injunction preventing me from handing over any more documents to the *Guardian* or *The New York Times*. With every turn, I was feeling more concerned about what was going to happen. A few days later, on 17 November, the same day Cadwalladr published a story in the *Guardian* about the threats she had received, I had a seizure on a London street, blacked out, and was taken to a hospital. The doctors said that the cause was unclear.

Soon after I was released from the hospital, I asked Allen if there was anything we could do to secure the information that I had against any effort to keep it away from the public. Was there a surefire way to protect against injunctions in Britain? She said no but then paused, pointing out that the one exception is inside the Houses of Parliament, where the ancient laws of parliamentary immunity shield MPs from injunctions or libel claims in the courts. Discussions of legal principles dating back to the 1600s seemed academic at first, but what Allen told me gave me an idea to take Alistair Carmichael up on his offer to help. Meeting with Carmichael at his office in Parliament, I told him that I was probably under surveillance and that I needed him to lock away some hard drives to safeguard evidence in the event I was not able to publish it. Carmichael agreed and told me that, should it come to it, he would do whatever was necessary to get this information out, even if it meant using his parliamentary immunity. I handed him several hard drives, and for the rest of the time before the stories broke, we kept key evidence in his safe.

I also helped him secure some remarkable recordings. Dr Emma Briant is a British professor and information warfare expert who came across several Cambridge Analytica executives during the course of her research into CA's work for NATO. Even as someone who hangs around in military propaganda circles, she had been shocked by her conversations with the firm and had begun recording them. Cadwalladr had introduced us because Briant needed help getting the same kinds of protections I was able to secure working through Carmichael at Parliament. I sat in Alistair's office as Briant played a recording of Nigel Oakes, the CEO of SCL Group, Cambridge Analytica's parent company. 'Hitler attacked the Jews, because he didn't have a problem with the Jews at all, but *the people* didn't like the Jews,' said Oakes. 'So he just leveraged an artificial enemy. Well, that's exactly what Trump did. He leveraged a Muslim.' Oakes's company was helping Trump do what Hitler did, but he seemed to find the whole thing amusing. In a separate clip of a discussion between Briant and Wigmore, the Leave.EU communications director also seemed to be interested in reviewing the strategic nature of the Nazis' communication campaigns. In the tape, Wigmore is recorded explaining, 'The propaganda machine of the Nazis, for instance – if you take away all the hideous horror and that kind of stuff, it was very clever, the way they managed to do what they did. In its pure marketing sense, you can see the logic of what they were saying, why they were saying it, and how they presented things, and the imagery . . . And looking at that now, in hindsight, having been on the sharp end of this [2016 EU referendum] campaign, you think, crikey, this is not new, and it's just – it's using the tools that you have at the time.' As we played the recordings, Carmichael just sat there in silence.

Finally, in February 2018, Allen and I were invited to a screening room in the ITN building, which coincidentally was across the street from Tamsin's office on Gray's Inn Road. I watched as Nix shifted in his seat in our pretend dining room, trying to cater to his guests' whims and desires. I watched as each sentence was spoken, each mistake made. It was insane. Here I was watching Nix in full form, admitting to some of the grotesque things Cambridge Analytica had done and was willing to do. Nix discussed how he had met Trump 'many times' during the 2016 campaign. Turnbull went further, revealing how Cambridge Analytica had set up the 'crooked Hillary' narrative. 'We just put information into the bloodstream [of] the internet and then watch it grow,' he said. 'And so this stuff infiltrates the online community, but with no branding, so it's unattributable, untrackable.' As I watched, I could barely contain myself. My experience was finally being validated by Nix's own words.

The footage was perfect. Nix and Turnbull were caught dead to rights, casually offering to find *kompromat* and to blackmail a Sri Lankan minister. Nix, draping one leg over the other and sipping a drink, said:

> Deep digging is interesting. But you know, equally effective can be just to go and speak to the incumbents and to offer them a deal that's too good to be true, and make sure that's video-recorded. You know, these sorts of tactics are very effective. Instantly having video evidence of corruption. Putting it on the internet, these sorts of things . . .

> We'll have a wealthy developer come in – somebody posing as a wealthy developer . . . They will offer a large amount of money to the candidate, to finance his campaign in exchange for land, for instance. We'll have the whole thing recorded on cameras. We'll blank out the face of our guy and then post it on the internet.

Yes, Nix actually proposed conducting a sting operation, right in the middle of ours. I sat there watching with Allen and the Channel 4 team, taking in the sheer irony of it all. And then Nix went on and offered to:

> send some girls around to the candidate's house. We have lots of history of things . . . We could bring some Ukrainians in on holiday with us, you know what I'm saying . . . They are very beautiful. I find that works very well . . . I'm just giving you examples of what can be done and what has been done . . . I mean, it sounds a dreadful thing to say, but these are things that don't necessarily need to be true, as long as they're believed.

After months of work and endless wrangling, we finally had all the elements in place. This Channel 4 footage would serve as the story's coup de grâce, and in that moment I finally felt confident that we were actually going to stop Cambridge Analytica.

AT LONG LAST, it was agreed that the print stories and the corresponding broadcast investigation would come out during the last two weeks of March 2018. A couple of weeks before publication, I met with Damian Collins, the chair of Parliament's Digital, Culture, Media and Sport Committee (DCMS), at his office in Portcullis House, a modern glass building on the Parliamentary Estate. Collins had opened up an official inquiry into social media disinformation, and several MPs and committee chairs I'd talked with had recommended that I meet with him. Collins was extremely polite and posh and spoke with the preppy charm that English Tories of a certain breed seem to have. I was impressed with him from the start. He was far more aware of what Cambridge Analytica was than any other MP I had met, and he had in fact already called Nix to testify several months prior. Nix had denied before the committee – on record – that Cambridge Analytica used any Facebook data. I told Collins that was false and that Nix may have misled the committee – which was quite serious, as it was potentially contempt of Parliament. I plugged one of the drives from Carmichael's safe in to my laptop and turned the screen to Collins. On the screen was a fully executed contract for Facebook data with both Nix's and Kogan's signatures signed in bright blue ink. We spent several hours going through internal documents from CA that established that the company used Facebook data and had relationships with Russian companies, and showing some of the extremely gruesome propaganda they had disseminated showing people being murdered. After Collins and the committee staff identified the documents they needed, I made a copy and handed a drive over to him. We agreed that, two weeks after the scheduled publication date, his inquiry would call me to testify in public. On that day, he would start a document dump via the committee of the documents I had given him.

At the same time all this was happening, I had been updating the Information Commissioner's Office – the government agency that investigates data crimes – on the evidence we were gathering about Cambridge Analytica's illicit activities. After seeing the Channel 4 footage, I told Commissioner Elizabeth Denham that CA was still at it, proposing to commit crimes on behalf of prospective clients. The ICO asked us to hold off on breaking the stories, because they wanted to conduct a raid before everything went public. They didn't want CA to have a chance to delete evidence. I gave them all the evidence I had, including copies of CA executives' files, project documents and internal emails, which they then passed along to the National Crime Agency, the British equivalent of the FBI. I had to curate the evidence, since it was all quite complicated, in order for the ICO to execute proper warrants for the raid. Tamsin and I were also preparing witness statements and a full written opinion to be provided to the Electoral Commission, about the crimes committed by the Leave campaign. We were hardly sleeping – working on legal documents, advising law enforcement, managing the journalists. It was an exhausting time. But, finally, everything was coming together.

About a week before publication, the *Guardian* sent out right-to-reply letters to the people and companies named in their reporting. The letters are a customary British journalism practice intended to give people a chance to respond to allegations before articles are published. On 14 March, I received a letter from Facebook's lawyers demanding that I hand over all my devices for their inspection, citing the Computer Fraud and Abuse Act and the California Penal Code in an attempt to intimidate me with criminal liability. On 17 March, the day before publication, Facebook threatened to sue the *Guardian* if it moved forward with the articles, insisting that there had been no data leak. When the company realised that publication was inevitable, in an attempt to get ahead of the story and shift focus, it announced that it was banning me, Kogan and Cambridge Analytica from using the platform. the *Guardian* and *The New York Times* were furious that Facebook was using the extra notice it had been given, in good faith, to attempt to undermine the story with its announcement.

On the evening of 17 March, the *Guardian* and the *Times* worked all night to rush through the publication of their stories. The *Times* headline read: 'How Trump Consultants Exploited the Facebook Data of Millions.' The *Guardian* editors chose a more dramatic headline: '"I Made Steve Bannon's Psychological Warfare Tool": Meet the Data War Whistleblower.' The stories instantly went viral, and that night Channel 4 began running its series, including the devastating sting exposing Nix. The channel also released an interview with the defeated 2016 Democratic presidential candidate, Hillary Clinton, who described the allegations about Cambridge Analytica as 'very disturbing'. In the interview, Clinton said, 'When you have a massive propaganda effort to prevent people from thinking straight, because they're being flooded with false information and . . . every search engine, every site they go into, is repeating these fabrications, then yes, it affected the thought processes of voters.' Cadwalladr's story blew up after that. Two other *Guardian* journalists, Emma Graham-Harrison and Sarah Donaldson, wrote articles explaining how it all connected. Their brilliant storytelling obviously resonated with regular non-tech folk and caused a massive jamboree on social media (save for Facebook, which instead promoted its own press release in its trending news stories section). The *Times* story focused on the

Facebook data breach, identifying it as 'one of the largest data leaks in the social network's history'. Reporters Matthew Rosenberg and Nicholas Confessore, bylined with Cadwalladr, also connected the dots between Bannon, Mercer and Cambridge Analytica and explained in detail how they had used Facebook data to propel Trump to victory.

In London, the British authorities had already been investigating both Cambridge Analytica and Facebook for months, as I had handed over evidence to them before the story broke. But while the UK Information Commissioner's Office was in the process of applying for warrants in the British courts to execute a search of Cambridge Analytica's offices and seize evidence, Facebook had already hired a 'digital forensics firm' to examine Cambridge Analytica's servers, beating the authorities into CA's headquarters. Although the ICO required a warrant to enter, Facebook did not, as it had been granted access by CA. When Facebook found out that the story was about to break, it contacted Cambridge Analytica, which agreed to provide Facebook access to its servers and computers while the ICO was still in the process of requesting a warrant. But when the ICO was tipped off that Facebook had entered CA's headquarters, they were furious. They had never seen a company take such brazen steps to handle evidence that would soon become the subject of a judicial search warrant. What made the situation even worse was that Facebook was not a mere bystander in this affair – Facebook's data was also a subject of the investigation and the company was inside a potential crime scene handling evidence that could affect its own legal liability. The ICO sent agents to the scene, escorted by the police. Late that night, a dramatic standoff ensued between ICO agents, British police and Facebook's 'forensic auditors.' Facebook's auditors were ordered to drop everything and immediately leave Cambridge Analytica's offices, and they agreed to stand down. Elizabeth Denham, the UK information commissioner, was so incensed by Facebook's actions that she made a rare appearance the next day on British news, issuing a statement that Facebook's actions would 'potentially compromise a regulatory investigation'.

The reaction on both sides of the Atlantic was instant and explosive. I was called before the parliamentary inquiry into 'fake news and disinformation'. It would be the first of many public and secret hearings, covering everything from Cambridge Analytica's use of hackers and bribes to Facebook's data breach to Russian intelligence operations. Mark D'Arcy, the BBC's parliamentary correspondent covering the hearing, said, 'I think the [DCMS committee] hearing with Chris Wylie is, by a distance, the most astounding thing I've seen in Parliament.'

In Washington, the Federal Trade Commission and the Securities and Exchange Commission launched investigations, while lawmakers in the United States and the UK began calling for Facebook's CEO, Mark Zuckerberg, to testify under oath. The Department of Justice and the FBI flew to Britain to meet me in person on a Royal Navy base a few weeks after the story broke. The NCA had borrowed the building from the Royal Navy.

As Facebook's stock slid, Zuckerberg remained out of sight. He finally emerged on 21 March, with a Facebook post saying he was 'working to understand exactly what happened' and saying there had been a 'breach of trust between Kogan, Cambridge Analytica and Facebook.' The hashtag #DeleteFacebook started trending on Twitter, with Elon Musk stoking the fire by tweeting that he'd deleted the Facebook pages of SpaceX and Tesla. As I prepared for my public testimonies, I listened to Cardi B, the American rapper who had released her debut album only a few weeks after the story broke. The record's (purely coincidental) title, *Invasion of Privacy,* quickly prompted memes to circulate on social media with Mark Zuckerberg's face appended to an edited version of the now-platinum album's cover. It began to look as if the story was tipping into the zeitgeist, and people who had already been feeling uneasy about how Facebook operates were now having their fears confirmed in the most public fashion. In the throes of this PR nightmare, Zuckerberg bought ad space in major newspapers to publish a letter of apology, only a couple of weeks after Facebook first threatened to sue the *Guardian* in an attempt to shut down the story, but the letter did little to quell the anger. Just two weeks later, he faced two long days of grilling by US congressional leaders.

In Britain, there was still more to report. This time it was about Brexit. As the stories were breaking in America, a new tranche of right-to-reply letters were sent out to those involved in Vote Leave.

Dom Cummings and Stephen Parkinson were among the recipients. It was only when Sanni arrived at our lawyer's office that evening after receiving a flurry of calls from former Vote Leave staffers that we realised what Parkinson had done in response. Parkinson had responded in the most cruelly personal way imaginable. At the time, Parkinson was working as a senior adviser to Prime Minister Theresa May, and a day before the *Guardian* published the story, the Downing Street press office issued an official statement, which we discovered only when *The New York Times* asked us to comment. In the statement, Parkinson revealed his relationship with Sanni and dismissed the accusations as bitterness over their breakup. Sanni is a Pakistani Muslim, and he had not yet come out to his family as gay, because it would have put his relatives back in Pakistan in physical danger, a fact that Parkinson knew full well. Despite this, he chose to out Sanni to the world's media and let his former intern deal with the consequences. This was the first time, at least in recent history, that the press office of the prime minister had publicly outed someone in an act of retaliation. When Sanni heard about the statement, he looked solemnly into everyone's eyes and sat back in his chair. Allen and Cadwalladr would eventually convince Cummings to remove a blog post he had written in response about the affair, but the damage was done. Parkinson had done exactly what he intended to.

The Vote Leave revelations had to compete with the cover of the Sunday edition of the *Daily Mail:* 'PM's Aide in Toxic Sex Row Over Pro-Brexit Cash Plot.' Continuing their vilification of LGBTQ people, Britain's right-wing press had reduced Sanni and his evidence of the largest breach of campaign finance law in British history to nothing more than 'toxic sex'. By now, Sanni's family in Karachi had to take security measures back at home due to the threat of violence that LGBTQ people and their families face in Pakistan. His life and the lives of the people he loved had been upended. I'll never forget watching Sanni through a window as he sat alone in Allen's office at half past midnight, dialling his mother to tell her that, yes, he was gay. It was a moment in which the courage and consequences of his decision to come out as a whistleblower were inseparable. In the following days, the violence Sanni faced only worsened as he was followed by people with hidden cameras, and photos of him and me inside a gay bar were later published on British alt-right sites with incredibly homophobic commentary. In Parliament, Prime Minister Theresa May herself defended the actions of Parkinson. It was heartbreaking to watch, but it made me so proud to call Sanni my friend.

ON THE EVENING OF 20 March, three days after the Cambridge Analytica story broke, I went to the Frontline Club, in London, with Allen and Sanni for my first public appearance. I was swarmed by photographers as I walked in, and the venue heaved with reporters from around the world. Journalists had grabbed the closest seat they could get. All along the back, there were cameras from more than twenty news channels, and with so many people crammed inside, it was getting hot. The journalist and privacy activist Peter Jukes interviewed me before the crowd, and I then took questions. When I couldn't take any more attention, I left via a discreet escape route. So as not to make a scene, the plan was for Allen to leave a few minutes later. Outside, I turned right and was headed down Norfolk Place when a man approached me. He held a glowing phone up to my face. I took a step backward, confused and a little alarmed. I asked what he wanted, and he told me to just look at his phone.

My eyes adjusted, and I could make out a screenshot of a Cambridge Analytica invoice to UKIP. He then swiped to what looked like an email from Andy Wigmore, the communications director of Leave.EU, to someone with a Russian name. I didn't have long to look at it, but it seemed the message was about gold. 'They worked with the Russians,' the man told me. At this point, Allen and the others came out with some of the others. When they saw the man, they were worried for my safety and ran up, trying to grab him. The man shook them off and got away. I was just in a daze from it all. Earlier that day I had been on back-to-back live TV interviews and had been chased down by photographers. It was an overwhelming day. In the car ride back to Allen's office to collect my bag, I told her that I wasn't really sure what it was about but that the messages looked real enough: I recognised the bank account details. Later that week, Allen received a cryptic message and called to say she thought that the man who had stopped me on the street was trying to contact me.

I had thought most of my work as a whistleblower was over by this point, but what transpired next led me to information so sensitive that my meetings with the House Intelligence Committee that day in June 2018 had to be conducted in the SCIF, underneath the United States Capitol. In the two months before that secret hearing, I met the man in several random locations scattered across London, and it became clear that he had access to files belonging to Leave.EU co-founder Arron Banks and communications director Andy Wigmore. The documents constituted a record of extensive communications between Leave.EU, a major alt-right pro-Brexit campaign, and the Russian embassy in London during the Brexit campaign. Once we were assured of their authenticity, Allen and I contacted MI5 and the National Crime Agency.

In April, Allen met with an NCA officer in one of their unmarked offices inside a major London train station to update them on my behalf, as we couldn't be sure that I wasn't being followed. We were both worried because we'd learned that the man was travelling with these documents, which potentially contained evidence of a Russian intelligence operation, throughout Ukraine and Eastern Europe. The NCA notified the British embassy in Kiev about the situation. Then we lost track of him and his phone was disconnected. We were all deeply concerned about his safety.

Several weeks later, the man reemerged and wanted to meet again. Allen and I decided that I would secretly record my meetings with the man. We handed over copies of the recordings and screenshots of the documents to the British authorities. We also notified the Americans, because we saw evidence that the Russians were speaking with Cambridge Analytica clients immediately before and after the clients met with the Trump campaign. We eventually had a meeting with California congressman Adam Schiff, then the ranking member of the House Intelligence Committee, in Nancy Pelosi's office in the Capitol. Allen and I told Congressman Schiff about the existence of these documents. I agreed to return to DC with the documents, which were kept secure in Carmichael's safe at Parliament.

Soon after this DC meeting, I was contacted by Fusion GPS, the private intelligence firm that had put together the Trump-Russia dossier authored by Christopher Steele. Steele's firm had learned about the documents and recordings I had through a British source, and Fusion GPS told us they had documents and information of their own that illuminated the same set of connections – between the Russians, Brexit and Trump's campaign. We all agreed to meet in the office of DCMS committee chair Damian Collins. Like a jigsaw, Collins, Fusion GPS and I had each acquired different sets of documents about the same events, and we began piecing everything together. Allen approached the NCA again, but they declined to take action, so we instead handed everything over to the US House Intelligence Committee, which agreed to pass it on to the appropriate American intelligence channels. If the UK authorities were not going to touch the evidence we had about Brexit and the Russian embassy on their own, we hoped that when the American agencies got access to the documents they would put pressure on their British colleagues to take action.

THE DOCUMENTS TELL A remarkable story. In 2015, not long after I left Cambridge Analytica, the UK Independence Party-backed Leave.EU campaign retained the firm to, as Leave.EU stated at its campaign launch, 'map the British electorate and what they believe in, enabling us to better engage with voters.' The CA-UKIP relationship was one that was fostered by Steve Bannon. Once Banks and Wigmore had been speaking to Bannon about Cambridge Analytica, Nigel Farage introduced them to his friend Robert Mercer. Mercer was keen to help their budding alt-right movement, but the American billionaire, like all foreigners, was legally barred from donating or substantively interfering in British political campaigns. So the Brexiteers were told by the billionaire that the data and services of Cambridge Analytica could be useful, and Bannon offered to help. Farage, Banks and company accepted Bannon's offer, consummating the emerging Anglo-American alt-right alliance with databases and algorithms.

It was this relationship that became a focal point of interest for the House Intelligence Committee, as it appeared that this relationship was exploited by the Russian embassy as a discreet vehicle into the Trump campaign. In November 2015, Leave.EU publicly launched its referendum campaign with Brittany Kaiser, who, in addition to working at Cambridge Analytica, was appointed as Leave.EU's new director of operations. On the campaign, Kaiser was to have a special focus on deploying CA's microtargeting algorithms.

Shortly before the public launch with Cambridge Analytica, the top donors to UKIP and Leave.EU – Arron Banks and Andy Wigmore – began their flirtations with the Russian government. It all started at the 2015 UKIP conference, in Doncaster, with a meeting between the two and Alexander Udod, a Russian diplomat who invited them to meet his ambassador at the Russian embassy. A few weeks later, after what was described in correspondence as a 'boozy six-hour lunch' with Alexander Vladimirovich Yakovenko, the Russian ambassador in London, Banks and Wigmore met the ambassador a second time and were given an enticing offer, which Banks then extended to several associates, including Jim Mellon, a prominent investor and Brexit backer. The Russian embassy was interested in facilitating introductions for some potentially lucrative deals to invest in what Banks referred to in an email as 'The Russian Gold Play'. The pitch to the men was made through the ambassador, who introduced them to Siman Povarenkin, a Russian businessman. Povarenkin suggested that several Russian gold and diamond mines were about to be consolidated and partially privatised. The embassy made clear that the deal would be backed by Sberbank, a Russian state bank, subject to US and EU sanctions. The advantage of working through the embassy and Sberbank, the UKIP donors were told, was that it 'leads to certain opportunities not available to others'.

In the time leading up to the announcement that Cambridge Analytica would be working on Leave.EU's campaign, the contact with the Russian embassy continued. In an email responding to a meeting invitation from a Russian official, Banks wrote the embassy to say, 'Thank you Andy and myself will be delighted to attend lunch to brief the Ambassador on the 6th November. There is massive interest in this referendum in the USA as well, and we are shortly visiting Washington to brief key [*sic*] on the campaign.' On 16 November, 2015, the day after the announcement, Banks and Wigmore were invited back to the embassy for more meetings. We do not know exactly what transpired in the embassy that day, but we do know that the Brexiteers then flew to America to meet with their Republican counterparts and that the Russian embassy was aware of these trips. We also know that Banks and Wigmore were keen on keeping Ambassador Yakovenko updated, saying in one text message to the ambassador in January 2016, 'Andy and myself would love to come and update you on the campaign. It's all happening. All the best, Arron.'

Why Banks would tell the Russian ambassador about his American political contacts or the Brexit referendum campaign if he was strictly dealing with the Russians on business is unclear, but the meetings certainly had an effect on the Brexiteers. In one chain of correspondence, one of the men discussed helping to create a Brexit-style movement in Ukraine, with the goal of fighting pro-EU narratives in a country Russia has long fought to keep inside its sphere of influence. They later decided against a foray into Ukraine and in one email even discussed whether a sentence in a draft of a press release might be seen as 'too overtly Russophile', but Wigmore nonetheless responded with the suggestion to 'send a note of support to the Ambassador'.

Banks and Wigmore kept in contact with the Russian embassy; Wigmore wrote to invite Russian diplomats to attend Leave.EU events – including their Brexit victory party in June 2016. Though Banks reportedly consulted experts about the Russian gold and diamond mine investment offers, he told reporters he ultimately turned all of them down. Wigmore had also decided 'not to proceed further' with an investment. But shortly after the Brexit campaign concluded, an investment fund associated with Jim Mellon, one of UKIP's major donors, reportedly made an investment in Alrosa, the Russian state-owned diamond company, which was partially privatised. However, a representative of the firm said that the specifics of the investment were made without Mellon's knowledge and that the fund had made an earlier investment in Alrosa when public shares were first offered in 2013. In late July 2016, a month after Brexit was won and just weeks after the Russian intelligence hack of the Democratic National Committee's files and emails was leaked, Alexander Nix went to a polo match and was photographed with Ambassador Yakovenko sharing a bottle of Russian vodka. Coincidentally, this was also around the time that Nix was seeking to get access to WikiLeaks's information for the Trump campaign.

With Brexit won, Farage and Banks set their sights on America, now deep in the midst of the 2016 campaign. Over the course of 2016, these Brits campaigned vigorously for Trump, with Farage attending a myriad of public events for the Republican candidate. It seemed normal to casual observers at the time that then-candidate Trump, declaring himself 'Mr Brexit', would invite the lead figures of UKIP to his rallies. But what many Americans don't understand is how connected the alt-right is. It is a coordinated global movement. And it became a massive security risk in 2016.

On 20 August, 2016, Sergey Fedichkin, the Third Secretary of the Russian embassy, was sent an email by Andy Wigmore, with the subject line 'Fwd Cottrell docs – Eyes Only'. There were a couple of attachments and a cryptic one-line message: 'Have fun with this.' The attachments contained legal documents pertaining to George Cottrell's arrest by US federal agents. At the time, Cottrell was the chief of staff to Nigel Farage and the head of fundraising for UKIP. Farage said later that he knew nothing of Cottrell's illegal activities. After flying to America to celebrate their recent Brexit victory before a large Trump rally at the 2016 Republican National Convention, Cottrell and Farage were at Chicago O'Hare airport, about to return to England. Before takeoff, several agents boarded the plane and arrested Cottrell on multiple counts of conspiracy to commit money laundering and wire fraud. He had also been linked to Moldindconbank, the Moldovan bank alleged to be a key player in the 'Russian Laundromat' money-laundering scheme. Wigmore is recorded, in emails I obtained from my contact, sending Russian diplomats copies of the US Justice Department's charges. Following a plea agreement, Cottrell pleaded guilty to wire fraud.

The Russian embassy clearly knew how tightly connected key figures in the Brexit movement were to the Trump campaign, and the embassy kept cultivating their relationship to the point where they received documents from Wigmore about their UKIP associate's FBI arrests. Why should Americans care about what Russia was doing in Britain? Because these Brexiteers shared the same data firm, in Cambridge Analytica, and the same adviser, in Steve Bannon, and they were clearly keeping the Russians informed at each step of the way. And these same Brexiteers were some of the very first people invited to Trump Tower after his surprise victory. The president-elect of the

United States met with British citizens who were regularly briefing the Russian government.

As the journalists celebrated having exposed Cambridge Analytica and plunged the stock of an intransigent Facebook, I did not feel joyous. I was numb. It felt like watching the death of someone whose time has come. It was the most gruelling and arduous thing I have ever been through. I only began to process what had happened months later, after the adrenaline subsided. I realised how much trauma I had endured and I allowed myself to feel the pain of the experience, a pain made all the more acute by the role I had played in this disaster. As I saw Trump rise to power and watched as he banned citizens of Muslim states from entering the United States and gave justifications for white supremacist movements, I couldn't help feeling that I had laid the seeds for this to happen. I had played with fire, and now I watched as the world was burning. In heading to Congress, I was not simply going there to give my testimony. I was attending my own confessional.

12

REVELATIONS

I WON'T TELL YOU WHERE I LIVE, EXACTLY. IT'S SOMEWHERE between Shoreditch and Dalston, in the East End of London. I am the pink-haired guy who lives on the top floor, but I don't really stand out much. The neighbourhood is working class in its roots, and many buildings here were once factories in London's industrial age. Faded paint on smoke-stained brickwork advertises long-gone products from a century ago. There is a détente between the Indian, Pakistani and Caribbean communities that moved here in the last wave of Commonwealth immigration and the new wave of artists, gays, students and grungy weirdos who are being pushed out of central London by the cost of living. There are art deco cinemas, roof gardens and the restless cacophony of intoxicated clubgoers drinking cans of Red Stripe until 4 a.m. every weekend. One often sees completely veiled Muslim women shopping in the same off-license greengrocer as tattoo-clad club kids with asymmetrical hair. It is still a place where I can walk outside in relative anonymity.

My building is old, built in a time before the internet was even imaginable and when indoor plumbing was still a novelty. The floor is wooden and solid, but every so often it creaks as you take a step. There are extra bolts on the door, installed after a group of men kept calling by the week after I went public. My neighbours started complaining, until they realised who I was. Now they let me know anytime they see people loitering nearby.

There are many things missing where I live. In my living room, there's a stand in the far corner where there used to be a television. Wires still dangle from the walls there. It was a smart TV that connected to my Netflix and social media accounts, and it had a microphone and camera. In my room, there is a nightstand with a drawer that is lined with a special metallic fabric that prevents any devices in the drawer from sending or receiving electronic signals. As part of my bedtime ritual, I leave my devices in there. Across the room in my closet are my old electronics from my life before. An unplugged Amazon Alexa sits alone, buried among a pile of other electronic rubbish – tablets, phones, a smartwatch – that I have yet to dispose of properly. In another box sit the remnants of hard drives, degaussed, smashed up, or acid-bleached after the evidence on them was handed over to the authorities. The data is gone forever, and I might as well throw them out, but I feel oddly sentimental about them.

In the living room, I have an antique wooden desk from an old factory, and on it sits an air-gapped laptop that has never been connected to the internet. I used it to work through evidence handed over to the House Intelligence Committee. In the drawer is the blank laptop I use for travelling, in case it is searched at the border. My personal computer sits in the living room, encrypted and locked down with a physical U2F key. The cameras are taped, although there is little you can do about the built-in microphone. On the floor, there is a private VPN server connected to the wall, which in turn connects onward onto other servers.

There is a security camera at the entrance of my building that relays data to a security company. I have no idea if any of it is encrypted, so who knows who is watching. When I leave my house, I bring a portable panic button, but I have not yet needed to use it. The NCA put me on a watch list connected to one of my phones. If I call, they will prioritise a response, even if I say nothing to the operator. My backpack always has a portable hardware VPN router in case I need to connect to insecure Wi-Fi, as well as several Faraday cases that I got in pink because it was cute. I often wear a hat, but people will still recognise me, even a year later. Almost daily, I get the question 'Are you … *the whistleblower*?'

My life now looks like that of a paranoid man, but after being assaulted in the street, receiving threats from rogue private security firms, having my hotel room broken in to late at night as I was sleeping, and experiencing two hacking attempts on my email in the past twelve months, it is only sensible to be cautious. When I had my flat checked for security risks, the TV was deemed a risk, as it could be used to watch or listen to me without my ever knowing. As we dismantled it, I smiled at the irony of a TV that watches you.

In the days leading up to the story's publication, when Facebook began sending me legal threats and escalated my case up to its deputy general counsel and vice president, my lawyers realised that the company saw my whistleblowing as a major threat to its business. Having experience on other hacking cases, my lawyers knew what companies backed into a corner were willing to do. But Facebook was different. They did not need to hack me; they could simply track me everywhere because of the apps on my phone – where I was, who my contacts were, who I was meeting.

I disposed of my phone, and my lawyers bought new clean phones that have never touched Facebook, Instagram or WhatsApp. The terms and conditions of Facebook's mobile app asked for microphone and camera access. Although the company is at pains to deny pulling user audio data *for targeted advertising,* there is nonetheless a technical permission sitting on our phones that allows access to audio capabilities. And I was not an average user: I was the company's biggest reputational threat at the time. At least in theory, audio could be activated, and my lawyers were concerned that the company could listen in on my conversations with them or the police. Facebook already had access to my photos and my camera, which put them in a position to not just listen to me but also to see where I was. Even if I was alone in the bathroom taking a shower, I wasn't *really* ever alone. If my phone was there, so was Facebook. There was no escape.

But getting rid of my phone was not going to be enough. My mom, dad and sisters all had to remove Facebook, Instagram and WhatsApp from their phones for the same reason. But Facebook also knew who all my friends were, they knew where we liked to go out, what we wrote about in messages, and they knew where we all lived. Even hanging out with my friends became a risk, as Facebook had access to their phones. If a friend took a photo, Facebook could access it, and its facial recognition algorithms could, at least in theory, detect my face in the photos sitting on other people's phones, even if they were strangers to me.

As I was getting rid of my old electronics, my friends joked that it was as if I was exorcising the demons inside the machines, and one friend even brought over some sage to burn *just in case*. A funny gesture, of course, but in a way it really was an exorcism. We now live in a world where there are invisible spirits made of code and data that have the power to watch us, listen to us and think about us. And I wanted these spectres gone from my life.

ON 16 MARCH, 2018, a day before the *Guardian* and *The New York Times* published my story, Facebook announced that it was banning me from not only Facebook but also Instagram. Facebook had refused to ban white supremacists, neo-Nazis and other armies of hate, but it chose to ban me. The company demanded that I hand over my phone and personal computer and said that the only way for me to be reinstated was, in effect, to give them the same information I was providing the authorities. Facebook behaved as if it were a nation-state, rather than a company. The firm did not seem to understand that I was not the subject of investigation – *they were*. My lawyers advised me to refuse their demands, so as not to interfere with a lawful police and regulatory investigation. Later, when I was working with the authorities, the ban made it far more difficult to hand over evidence that was sitting in my Facebook account, and the investigation into what happened during the Brexit referendum suffered as a result.

They say you appreciate something only when it's gone, and it was only when I was erased from Facebook that I truly realised how frequently my life touched their platform. Several of my phone's apps stopped working – a dating app, a taxi app, a messaging app – because they used Facebook authentication. Subscriptions and accounts I had on websites failed for the same reason. People often talk about a dualism: the cyber world and our 'real lives'. But after having most of my digital identity confiscated, I can tell you they are not separate. When you are erased from social media, you lose touch with people. I stopped getting invited to parties – not intentionally, but because those invites always happened on Facebook or were posted on Instagram. Friends who did not have my new phone number found it nearly impossible to get hold of me, except by trying to send an email to my lawyers. When I got through the thick of the whistleblowing, it would only be in coincidental encounters at clubs or bars that I would make contact with people I had not seen in months.

And now, when guys on dating apps ask to check out my Instagram profile, it starts an awkward explanation about how I was banned – and that I'm not catfishing, I promise. It's as if my identity has been confiscated and people no longer believe that I am who I say I am. Sometimes I get recognised as *that guy*, and people worry that someone might start watching them if they decide to meet me. I always tell them that they needn't worry, because these companies are already tracking them 24/7. This ban was nothing more than a dick move by Facebook, and it felt like trolling by frightened bullies. For me, it created at most an annoying personal hassle and was not nearly as consequential to my life as the kinds of retaliation that other whistleblowers have experienced. (Not to mention the degree of damage to modern society that the platform had already aided and abetted.) But it showed me just how integral my online identity had become to so many facets of my life – and that my identity was afforded no due process rights or an impartial adjudication. Four days after my ban, during an emergency debate in Parliament, the British secretary of state for culture said that Facebook's ability to unilaterally ban whistleblowers was 'shocking', because it raised serious questions about whether a company should be able to wield this kind of unchecked power.

Hundreds of millions of Americans have entered into Facebook's invisible architecture thinking it was an innocuous place to share pics and follow their favourite celebrities. They were drawn into the convenience of connecting with friends and the ability to fend off boredom with games and apps. Users were told by Facebook that the enterprise was about bringing people together. But Facebook's 'community' was building separate neighbourhoods just for *people who look like them*. As the platform watched them, read their posts, and studied how they interacted with their friends, its algorithms would then make decisions about how to classify users into digital neighbourhoods of *their kind* – what Facebook called their 'Lookalikes'. The reason for this, of course, was to allow advertisers to target these homogeneous Lookalikes with separate narratives just for people of their kind. Most users would not know their classification, as the other neighbourhoods of people who did not look like them would remain unseen. The segmentation of Lookalikes, not surprisingly, pushed fellow citizens further and further apart. It created the atmosphere we are all living in now.

As the birthplace of social media, America was eased into the new digital commons of newsfeeds, followers, likes and shares. And, as with the incremental effects of climate change on our shorelines, forests and wildlife, it can be hard to fully picture the scale of change of something that envelops us. But there are cases where we can see the stark effects of social media, cases where it suddenly hits a country in full force. In the mid-2010s, Facebook entered into Myanmar and grew rapidly, quickly reaching 20 million users in a country of 53 million people. Facebook's app came preinstalled on many smartphones sold in the country, and market research identified the site as one of the primary sources of news for Burmese citizens.

In August 2017, hate speech surged on Facebook targeting the Rohingya, a predominantly Muslim minority group in Myanmar, with narratives of a 'Muslim-free' Myanmar and calls for ethnic cleansing of the region going viral. Much of this was propaganda created and disseminated by military personnel conducting information operations. After Rohingya militants launched a coordinated attack on the police, the Burmese military capitalised on a surge in support they received online and proceeded to systematically kill, rape and maim tens of thousands of Rohingya. Other groups joined in the slaughter, and calls to action to murder Rohingya continued to go out on Facebook. Rohingya villages were burned and more than 700,000 Rohingya refugees were forced across the border into Bangladesh. Facebook was warned repeatedly by international and local organisations about the situation in Myanmar. The company banned a Rohingya resistance group from the platform but left the military and pro-government groups on the site, which enabled them to continue spreading hate propaganda. This was despite what United Nations officials called a 'textbook example of ethnic cleansing'.

In March 2018, the UN concluded that Facebook had played a

'determining role' in the ethnic cleansing of the Rohingya people. Violence was enabled by Facebook's frictionless architecture, propelling hate speech through a population at a velocity previously unimaginable. Facebook's apathetic response was positively Orwellian. 'There is no place for hate speech or content that promotes violence on Facebook, and we work hard to keep it off our platform,' read Facebook's statement about its facilitating role in the ethnic cleansing of forty thousand human beings. It seemed for all the world that if you wanted to maintain an oppressive regime, Facebook would be an excellent company to turn to.

What was supposed to be so brilliant about the internet was that people would suddenly be able to erode all those barriers and talk to anyone, anywhere. But what actually happened was an amplification of the same trends that took hold of a country's physical spaces. People spend hours on social media, following people like them, reading news articles 'curated' for them by algorithms whose only morality is click-through rates – articles that do nothing but reinforce a unidimensional point of view and take users to extremes to keep them clicking. What we're seeing is a *cognitive segregation,* where people exist in their own informational ghettos. We are seeing the segregation of our realities. If Facebook is a 'community', it is a gated one.

Shared experience is the fundamental basis for solidarity among citizens in a modern pluralistic democracy, and the story of the civil rights movement is, in part, the story of being able to share space together: being in the same part of the movie theatre or using the same water fountain or bathroom. Segregation in America has always manifested itself in insidiously mundane ways – through separate bus seats, water fountains, schools, theatre tickets and park benches. And perhaps now on social media. For Rosa Parks, being ordered to give up her bus seat was just one of the countless ways white America systematically ensured that her dark skin was separated and unseen – that she remained *the other,* not part of *their* America. And although we no longer allow buildings to segregate their entrances based on a guest's race, segregation rests at the heart of the architectures of the internet.

From social isolation comes the raw material of both conspiracism and populism: *mistrust*. Cambridge Analytica was the inevitable product of this balkanised cyberspace. The company was able to get its targets addicted to rage only because there was nothing to prevent it from doing so – and so, unimpeded, the company drowned them in a maelstrom of disinformation, with predictably disastrous results. But simply stopping CA is not enough. Our newfound crisis of perception will only continue to worsen until we address the underlying architectures that got us here. And the consequences of inaction would be dire. The destruction of mutual experience is the essential first step to *othering*, to denying another perspective on what it means to be *one of us*.

Steve Bannon recognised that the 'virtual' worlds of the internet are so much more real than most people realise. People check their phones on average fifty-two times per day. Many now sleep with their phones charging beside them – they sleep with their phones more than they sleep with people. The first and last thing they see in their waking hours is a *screen*. And what people see on that screen can motivate them to commit acts of hatred and, in some cases, acts of extreme violence. There is no such thing as 'just online' anymore, and online information – or disinformation – that engages its targets can lead to horrific tragedies. In response, Facebook, like the NRA, evades its moral responsibility by invoking the same kind of 'Guns don't kill people' argument. They throw up their hands and claim they can't control how their users abuse their products, even when mass murder results. *If ethnic cleansing is not enough for them to act, what is?* When Facebook goes on yet another apology tour, loudly professing that 'we will try harder', its empty rhetoric is nothing more than the *thoughts and prayers* of a technology company content to profit from a status quo of inaction. For Facebook, the lives of victims have become an externality of their continued quest to *move fast and break things*.

When I came out as a whistleblower, the alt-right's digital rage machine turned its sights to me. In London, enraged Brexiteers pushed me into oncoming traffic. I was followed around by alt-right stalkers and had photos of me at clubs with my friends published on alt-right websites with information about where to find me. When it came time to testify at the European Parliament, conspiracies about Facebook's critics were beginning to percolate through forums of the alt-right. As I testified, there were chants of *'Soros, Soros, Soros'* in the back. As I was leaving the European Parliament, a man came up to me on the street, shouting *'Jew money!'* At the time, these narratives seemed to come out of nowhere. Later, it emerged that Facebook, in a panic about its PR crisis, had hired the secret communications firm Definers Public Affairs, which subsequently leaked out fake narratives filled with anti-Semitic tropes about its critics being part of a George Soros-funded conspiracy. Rumours were seeded on the internet and, as I discovered personally, its targets took it as a cue to *take matters into their own hands*.

IN FEBRUARY 2013, a Russian military general named Valery Gerasimov wrote an article challenging the prevailing notions of warfare. Gerasimov, who was Russia's chief of the general staff (roughly equivalent to chairman of the US Joint Chiefs of Staff), penned his thoughts in the *Military-Industrial Kurier* under the title 'The Value of Science is in the Foresight' – a set of ideas that some would later dub the Gerasimov Doctrine. Gerasimov wrote that the '"rules of war" have changed' and that 'the role of nonmilitary means of achieving political and strategic goals has grown'. He addressed the uses of artificial intelligence and information in warfare: 'The information space,' he wrote, 'opens wide asymmetrical possibilities for reducing the fighting potential of the enemy.' Essentially, Gerasimov took the lessons of the Arab Spring uprisings, which were propelled by information sharing on social media, and urged military strategists to adapt them. 'It would be easiest of all to say that the events of the "Arab Spring" are not war, and so there are no lessons for us – military men – to learn. But maybe the opposite is true – that precisely these events are typical of warfare in the twenty-first century.'

Gerasimov's article was followed by another Russian military strategy paper, this one written by Colonel S. G. Chekinov and Lieutenant General S. A. Bogdanov. Their paper took Gerasimov's idea even further: the authors wrote that it would be possible to attack an adversary by 'obtain[ing] information to engage in propaganda from servers of the Facebook and [T]witter public networks' and that, with these 'powerful information technologies at its disposal, the aggressor will make an effort to involve all public institutions in the country it intends to attack, primarily the mass media and religious organisations, cultural institutions, non-governmental organisations, public movements financed from abroad and scholars engaged in research on foreign grants.' At the time, it was a radical new idea. Read today, it is a precise blueprint for Russia's interference in the 2016 election.

The history of warfare is the history of new inventions and strategies, many of which were born out of necessity. By most metrics, Russia's military is significantly weaker than that of the United States. The US military budget, at $716 billion, is more than ten times that of Russia. The United States has 1.28 million active military personnel, as compared with Russia's 1 million; has more than 13,000 total aircraft, as compared with Russia's 4,000; and has twenty aircraft carriers, whereas Russia has one. By all existing conventional measures, Moscow would never again be competitive with the United States in terms of 'great powers' warfare, and Vladimir Putin knew it. So the Russians had to devise another way to regain the advantage – one that had nothing to do with the physical battlespace.

It's difficult for military strategists to envision new forms of battle when they're focused on those at hand. Before the advent of flight, military commanders cared only about how to wage combat on land or at sea. It wasn't until 1915, when the French pilot Roland Garros flew a plane jerry-rigged with a machine gun, that military strategists realised that war could actually be waged from the skies. Then, once aircraft began engaging in attacks, army units on the ground pivoted as well, creating compact, rapid-fire antiaircraft guns. And so the evolution of war continued.

Information warfare has evolved in similar fashion. At first, no one could have imagined that Facebook or Twitter could be battlefield tools; warfare was waged on the ground, in the air, at sea and potentially in space. But the fifth domain – cyberspace – has proved to be a fruitful battleground for those who had the imagination and foresight to envision using social media for information warfare. You can draw a straight line from the groundwork laid by Gerasimov, Chekinov and Bogdanov, right through the actions of Cambridge Analytica, to the victories of the Brexit and Trump campaigns. In only five or so years, the Russian military and state have managed to develop the first devastatingly effective new weapon of the twenty-first century.

They knew it would work, because companies such as Facebook would never take the 'un-American' step of reining in their users. So Russia didn't have to disseminate propaganda. They could just get the Americans to do it themselves, by clicking, liking and sharing. Americans on Facebook did the Russians' work for them, laundering their propaganda through the First Amendment.

But this new era of scaled disinformation is not confined to the realm of politics. Companies like Starbucks, Nike and other fashion brands have found themselves targets of Russian-sponsored disinformation operations. When brands make statements that wade into existing social or racial tensions, there have been several identified instances in which Russian-sponsored fake news sites, botnets and social media operations have activated to weaponise these narratives and provoke social conflict. In August 2016, the football player Colin Kaepernick refused to stand for the American national anthem to protest systemic racism and police brutality toward African Americans and other minorities in the United States. The fashion brand Nike, Kaepernick's sponsor, stood behind the athlete, and a controversy ensued about Nike's response. But unknown to many at the time, Russian-linked social media accounts began to spread and amplify existing hashtags promoting a Nike boycott within hours of the scandal emerging. Some of this Russian-amplified content eventually made it into mainstream news, which helped legitimise the Nike boycott narrative as a purely homegrown protest. Cybersecurity firms also identified fake Nike coupons originating from alt-right groups that targeted African American social media users with offers like '75% off all shoes for people of colour'. The coupons were intended to create scenarios in which unwitting African American customers would try to use the coupons in a Nike store, where they would be refused. In the age of viral videos, this scenario could in turn create 'real' footage showcasing a racist trope of an 'angry black man' demanding free stuff in a store. So why would these disinformation operations target a fashion company and attempt to weaponise its brand? Because the objective of this hostile propaganda is not simply to interfere with our politics, or even to damage our companies. The objective is to tear apart our social fabric. They want us to hate one another. And that division can hit so much harder when these narratives contaminate the things we care about in our everyday lives – the clothes we wear, the sports we watch, the music we listen to or even the coffee we drink.

We are all vulnerable to manipulation. We make judgments based on the information available to us, but we are all susceptible to manipulation when our access to that information becomes mediated. Over time, our biases can become amplified without our even realising it. Many of us forget that what we see in our newsfeeds and our search engines is already moderated by algorithms whose sole motivation is to select what will engage us, not inform us. With most reputable news sources now behind paywalls, we are already seeing information inch toward becoming a luxury product in a marketplace where fake news is always free.

In the last economic revolution, industrial capitalism sought to exploit the natural world around us. It is only with the advent of climate change that we are now coming to terms with its ecological externalities. But in this next iteration of capitalism, the raw materials are no longer oil or minerals but rather commodified attention and behaviour. In this new economy of surveillance capitalism, *we are the raw materials*. What this means is that there is a new economic incentive to create substantial informational asymmetries between platforms and users. In order to be able to convert user behaviour into profit, platforms need to know everything about their users' behaviour, while their users know nothing of the platform's behaviour. As Cambridge Analytica discovered, this becomes the perfect environment to incubate propaganda.

With the advent of home automation hubs such as Amazon Alexa and Google Home, we are seeing the first step toward the eventual integration of cyberspace with our temporal physical reality. Fifth-generation (5G) mobile and next-generation Wi-Fi are already being rolled out, laying the foundations for the 'Internet of Things' (IoT) to become the new norm, where household appliances big and small will become connected to high-speed and ubiquitous internet networks. These mundane devices, whether they are a refrigerator, a toothbrush, or a mirror, are envisaged to use sensors to begin tracking users' behaviour inside their own homes, relaying the data back to service providers. Amazon, Google and Facebook have already applied for patents to create 'networked homes' that integrate in-home IoT sensors with online marketplaces, ad networks and social profiles. In this future, Amazon will know when you pop an aspirin, and Facebook will watch your kids play in the living room.

Fully integrated with intelligent information networks, this new environment will be able to watch us, think about us, judge us and seek to influence us by mediating our access to information – where 'it' can see us, but we cannot see 'it.' For the first time in human history, we will immerse ourselves in *motivated spaces* influenced by these silicon spirits of our making. No longer will our environment be passive or benign; it will have intentions, opinions and agendas. No longer will our homes be a sanctuary from the outside world, for an ambient presence will persist throughout each connected room. We are creating a future where our homes will think about us. Where our cars and offices will judge us. Where doors become the doormen. Where we have created the demons and angels of the future.

This is the dream that Silicon Valley has for us all – to surround us at every minute and everywhere. In Cambridge Analytica's quest for informational dominance, it was never going to be satisfied with just social data sets and had already begun to build relationships with satellite and digital TV providers. After tapping into connected televisions, Cambridge Analytica planned to find a way to integrate with sensors and smart devices in people's homes. Imagine a future where a company like Cambridge Analytica could edit your television, talk to your children and whisper to you in your sleep.

THE FOUNDATION OF OUR legal system is contingent upon the notion that our environment is passive and inanimate. The world surrounding us may *passively* influence our decisions, but such influence is *not motivated*. Nature or the heavens do not *choose* to influence us. Over centuries, the law has developed several fundamental presumptions about human nature. The most important of these is the notion of human agency as an irrefutable presumption in the law – that humans have the capacity to make rational and independent choices on their own accord. It follows that the world does not make decisions for humans, but that humans make decisions inside of that world.

This notion of human agency serves as the philosophical basis for criminal culpability, and we punish transgressors of the law on the grounds that they made a condemnable choice. A burning building may indeed harm people, but the law does not punish that building, as it has no agency. And so human laws regulate human acts, and not the motivations or behaviours of their surroundings. The corollaries to this are the fundamental rights we have. During the Enlightenment, the fundamental rights of people were articulated as core entitlements to protect the *exercise of human agency*. The rights to life, liberty, association, speech, vote and conscience are all underpinned with a *presumption of agency*, as they are outputs of that agency. But agency itself has not been articulated as a right per se, as it has always been presumed to exist simply by virtue of our personhood. As such, we do not have an express *right to agency* that is *contra mundum* – that is, a right to agency that is exercisable *against the environment itself*. We do not have a right against the heavens or the undue influence of motivated and thinking spaces to mediate the exercise of our agency. At the time of America's founding, a situation where our agency could be manipulated by a motivated and thinking environment was never contemplated as a possibility. For the Founding Fathers, this would have been a power known only to God.

We can already see how algorithms competing to maximise our attention have the capacity to not only transform cultures but redefine the experience of existence. Algorithmically reinforced 'engagement' lies at the heart of our outrage politics, call-out culture, selfie-induced vanity, tech addiction and eroding mental well-being. Targeted users are soaked in content to keep them clicking. We like to think of ourselves as immune from influence or our cognitive biases, because we want to feel like we are in control, but industries like alcohol, tobacco, fast food and gaming all know we are creatures that are subject to cognitive and emotional vulnerabilities. And tech has caught on to this with its research into 'user experience', 'gamification', 'growth hacking', and 'engagement' by activating ludic loops and reinforcement schedules in the same way slot machines do. So far, this gamification has been contained to social media and digital platforms, but what will happen as we further integrate our lives with networked information architectures designed to exploit evolutionary flaws in our cognition? Do we really want to live in a 'gamified' environment that engineers our obsessions and plays with our lives as if we are inside its game?

The underlying ideology within social media is not to enhance choice or agency, but rather to narrow, filter and *reduce* choice to benefit creators and advertisers. Social media herds the citizenry into surveilled spaces where the architects can track and classify them and use this understanding to influence their behaviour. If democracy and capitalism are based on accessible information and free choice, what we are witnessing is their subversion from the inside.

We risk creating a society obsessive about remembering, and we may have overlooked the value of forgetting, moving on or being unknown. Human growth requires private sanctuaries and free spaces where we can experiment, play, dabble, keep secrets, transgress taboos, break our promises and contemplate our future selves without consequence to our public lives until we decide to change in public. History shows us that personal and social liberation begins in private. We cannot move on from our childhoods, past relationships, mistakes, old perspectives, old bodies, or former prejudices if we are not in control of our privacy and personal development. We cannot be free to choose if our choices are monitored and filtered for us. We cannot grow and change if we are shackled to who we once were or who we thought we were or how we once presented ourselves. If we exist in an environment that always watches, remembers and labels us, according to conditions or values outside our control or awareness, then our data selves may shackle us to histories that we prefer to move on from. Privacy is the very essence of our power to decide who and how we want to be. *Privacy is not about hiding – privacy is about human growth and agency.*

But this is not merely about privacy or consent. This is about who gets to influence our truths and the truths of those around us. This is about the architectures of manipulation we are constructing around our society. And herein lies the lesson of Cambridge Analytica. To understand the harms of social media, we have to first understand *what it is.* Facebook may call itself a 'community' to its users, or a 'platform' to regulators, but it is not a service, in the same way a building is not a service. Even if you don't understand exactly how cyberspace works, it is important to understand that it now surrounds you. Every connected device and computer is part of an interconnected information architecture – and shapes your experience of the world. The most common job titles in most Silicon Valley companies are *engineer* and *architect*, not *service manager* or *client relations*. But unlike engineering in other sectors, tech companies do not have to perform safety tests to conform to any building codes before releasing their products. Instead, platforms are allowed to adopt *dark pattern designs* that deliberately mislead users into continual use and giving up more data. Tech engineers intentionally design confounding mazes on their platforms that keep people moving deeper and deeper into these architectures, without any clear exit. And when people keep clicking their way through their maze, these architects delight in the increase in 'engagement'.

Social media and internet platforms are not services; they are architectures and infrastructures. By labelling their architectures as 'services', they are trying to make responsibility lie with the consumer, through their 'consent'. But in no other sector do we burden consumers in this way. Airline passengers are not asked to 'accept' the engineering of planes, hotel guests are not asked to 'accept' the number of exits in the building, and people are not asked to 'accept' the purity levels of their drinking water. And as a former club kid, I can tell you that when bars or concerts are over capacity and heaving with ravers, fire inspectors will order those *consenting customers* to leave a building if the conditions become manifestly unsafe.

Facebook may say: If you don't like it, don't use it. But there are no comparable alternatives to the dominant players on the internet, just as there are no alternatives to electric, telecommunications or water companies. To reject the use of platforms like Google, Facebook, LinkedIn and Amazon would be to remove oneself from modern society. How are you going to get a job? How are you going to get information? How are you going to socialise with people? These companies love to talk about consumer choice, when they know that they have done everything in their power to become a necessary part of most people's lives. Getting users to click 'accept' after presenting them with a novella's worth of dense legalese (almost twelve thousand words in Facebook's case) is nothing but *consent-washing*. These platforms are purpose-built to run user consent through a blender. No one opts out of these platforms, because users have no other choice but to accept.

When Facebook banned me, they did not simply deactivate my account; they erased my entire presence on Facebook and Instagram. When my friends tried to look up old messages I had sent, nothing came up: my name, my words – everything – had disappeared. I became a shadow. Banishment is an ancient punishment to rid a society of its criminals, heretics and political radicals who jeopardised the power of the state or church. In ancient Athens, people could be banished from society for ten years for any reason with no opportunity for appeal. In the Stalinist period of the Soviet Union, *enemies of the state* would not just disappear; all remnants of their existence – photos, letters, news references – would be erased and cleansed from the annals of official history. Throughout history, the powerful have used social memory and collective forgetting as a powerful weapon to crush dissent and correct their preferred histories to shape the realities of the present. And if we want to understand why these technology companies behave this way, we should listen to the words of those who built them. Peter Thiel, the venture capitalist behind Facebook, Palantir and PayPal, spoke at length about how he no longer believes 'that freedom and democracy are compatible'. And in elaborating his views on technology companies, he expounded on how CEOs are the new monarchs in a techno-feudal system of governance. We just don't call them monarchies in public, he said, because 'anything that's not democracy makes people uncomfortable'.

The philosophical basis of authoritarianism rests in the creation of total certainty within society. The politics of certainty repositions the notion of freedom, where *freedoms from* replace *freedoms to*. Strict rules and laws are coercively enforced to govern and shape the behaviour, thoughts and actions of the polity. And the first tool of authoritarian regimes is always informational control – both in the gathering of information on the public through surveillance and the filtration of information to the public through owned media. In its early days, the internet seemed to pose a challenge to authoritarian regimes, but with the advent of social media, we are watching the construction of architectures that fulfill the needs of every authoritarian regime: surveillance and information control. Authoritarian movements are possible only when the general public becomes habituated to – and numbed by – a new normal.

THE INTERNET HAS FRUSTRATED these old assumptions about the law and the polity that it governs. The internet is both everywhere and nowhere – it is physically dependent on servers and cables, but it exists without a single location of primary residence. This means that a single digital act could partially occur in countless physical locations simultaneously, or an action in one place could result in effects in another place. This is because the internet is a type of *hyperobject* – like our climate and biosphere, the internet surrounds us and we live within it. The tech community often call their platforms 'digital ecosystems', with an implicit recognition that their construction is a digital container or realm for at least part of our lives to exist within. We cannot see it or touch it, but we know it exists around us by its effects.

Often I encountered police investigators unfamiliar with data crime using false analogies about finding the 'murder weapon', the 'location of the body', and linear 'chains of causation'. But data crimes are crimes that usually don't happen in one specific place. Data crime can often behave like pollution – it's everywhere generally, but nowhere specifically. Data is completely fungible and intangible, as it is merely a representation of information. It can be stored simultaneously in distributed servers around the world; where even when it's in a place, it's never entirely in that place. Servers based in country A handling data subjects in country B could be accessed by a person in country C and deployed on a platform in country D after receiving instructions from a company in country E with financing from country F. This was the nature of Cambridge Analytica's complex setup. Even if serious harms were clearly incurred, such as hacking, data theft, menacing threats or deception, it would be unclear who could be held responsible, and our known systems for assessing culpability were entirely incapable of the job.

We like to imagine our government as the captain of the ship, but when the ocean itself changes, our captains may find themselves unprepared and unable to navigate. In July 2018, Britain's Electoral Commission found that the Vote Leave campaign had broken the law, illegally coordinating with BeLeave. On 30 March, 2019 – one year after the Brexit whistleblowing stories broke – the Vote Leave campaign officially dropped its appeal of the EC's findings and fines, essentially admitting to what it had done. Some have asked, 'why should we care so much about a mere £700,000?' Let's be clear on this point: *Vote Leave's scheme was the largest known breach of campaign finance law in British history.* But even if it wasn't, elections, like a 100-metre sprint in the Olympics, are zero-sum games, where the winner takes all. Whoever comes first, even if it's by just a few votes or milliseconds, wins the whole race. They get to sit in the public office. They get the gold medal. They get to name your Supreme Court justices. They get to take your country out of the European Union.

The only difference, of course, is that if you are caught cheating in the Olympics, you get disqualified and lose your medal. There are no discussions of whether the doped athlete 'would have won anyway' – the integrity of the sport demands a clean race. But in politics, we do not presume integrity as a necessary prerequisite to our democracy. There are harsher punishments for athletes who cheat in sport than for campaigns that cheat in elections. Though they won by only 3.78 per cent, the Brexiteers claimed the entire 'will of the people' for themselves – and even when Trump *lost* the popular vote by 2.1 per cent, he too claimed victory. Despite proven cheating, Vote Leave did not have its Brexit medal taken away. No one was disqualified from running in future campaigns, and Vote Leave's two leaders, Boris Johnson and Michael Gove, were both allowed to run for prime minister. Crimes waged against our democracy were not considered by the political class to be 'real crime'. Many framed these transgressions as being on par with a parking fine, despite the very real harm we face when our civic institutions can be so easily undermined by criminals and hostile foreign states seeking to wage electoral terrorism on our society. And, of course, the most powerful people in Britain and America took the position that these crimes didn't even happen – rather, they were a 'hoax', the invention of the bitter opponents they had vanquished. This, in the face of what were once known as 'facts' and 'reality'.

You'd think that after pulling off a conspiracy to hack a world leader's private emails and medical records, bribe ministers, blackmail targets and shower voters with menacing videos of gruesome murders and threats, there would be some kind of legal consequence. But there were no consequences for anyone involved in Cambridge Analytica's African projects. It was too difficult to establish *jurisdictionality* – whether or not 'enough' of the crime happened in Britain to warrant prosecution in the English courts. Their servers were all over the world, the meetings happened in different countries, the hackers were based in yet another country, and Cambridge Analytica only *received* the hacked material in London but did not *request* the hacked material in the UK. Even though there were several witnesses to what happened, Cambridge Analytica simply got away with it. In fact, one of the managers from the Nigeria project eventually moved on to work in a senior position at the UK Cabinet Office on foreign affairs projects, sitting in the highest levels of the British government.

In America there were no consequences for Cambridge Analytica, either. The company knowingly and wilfully violated the Foreign Agents Registration Act. It conducted operations to suppress African American voters. It defrauded Facebook users and menaced them with disgusting content. It exposed hundreds of millions of private records of American citizens to hostile foreign states. And yet nothing happened, because Cambridge Analytica was set up for jurisdictional arbitrage. Tax evasion frequently involves setting up shell companies on tropical islands all around the world in an attempt to launder money through a complex enough chain of countries and companies, each with its own unique rules, that authorities lose track of where the money is. This is possible because money, like data, is a completely fungible asset and can be instantly moved through a global finance system. What Cambridge Analytica did was use complex corporate setups across jurisdictions not only to launder money but to launder something that was becoming just as valuable: *your data*.

In Britain, there were no consequences for AIQ either. After Sanni and I revealed proof of Vote Leave's unlawful scheme to overspend through AIQ and use it as a hidden proxy for Cambridge Analytica's targeting capabilities, it became the elephant in the room for the Brexit debate. Britain had already formally submitted its notice to exit the European Union. The notion that the wafer-thin Brexit result could have been affected by systematic cheating, data breaches and foreign interference was wilfully ignored because the ramifications were unimaginable. If the same events had happened in Kenya or Nigeria, there would have been prompt calls from British observers to hold a new vote.

Other British institutions failed as well. BBC executives, who were briefed by the *Guardian* on the story and given the entire corpus of evidence weeks in advance of publication, decided to drop the story days before it went public – it would have been too controversial. Instead the BBC interviewed Alexander Nix before the Channel 4 exposé was aired and did not include any comment from the whistle-blower. When I later appeared on *Newsnight*, the BBC's flagship evening news program, the host was at pains to keep interjecting that Vote Leave's law-breaking, which involved using unlawful money for billions of targeted Facebook ads, was merely another one of my 'allegations'. This was despite the fact that this was already an established finding of the Electoral Commission. Frustrated and confused, I then got into an argument about the meaning of a 'fact' and how bizarre it was that even with the published rulings of British legal authorities, the BBC still would not let me say that Vote Leave broke the law or that unlawful activity occurred on Facebook's watch.

The NCA suddenly dropped its investigation of Russian interference, even after it received evidence of the Russian embassy's dealings with Leave.EU. Later, the prime minister refused to deny that she had halted the investigation into Brexit. There was no parliamentary inquiry launched into the cheating that occurred during the Brexit referendum, and I ended up spending more time answering questions about Brexit in testimony before the United States Congress than in the British Parliament. Despite the lack of an inquiry in Britain, the

Canadian Parliament opened its own inquiry into AIQ's role in Brexit, to help the UK authorities compel answers from AIQ after the firm had successfully avoided its jurisdiction by remaining in Canada.

It turns out cheating is a pretty good strategy to win, as there are very few consequences. The Electoral Commission later conceded that even if the vote was won with the benefit of illegal data or illegal financing, the result still stands. Facebook refused to hand over the full details of what happened on its platform during Brexit or the number or types of voters who were profiled and targeted by illegal campaigns. Mark Zuckerberg defied three requests to testify before the British Parliament, and when fifteen national parliaments, collectively representing almost one billion citizens across six continents, banded together in a joint request to interview Zuckerberg, even over the phone, he still turned them down – twice. It seemed that Zuckerberg's time was more valuable than that of legislatures representing almost one seventh of the human race. Facebook learned that, despite the wrath of the media storm, there were actually very few consequences for simply ignoring the parliaments of the world – the company learned that it could behave like a sovereign state, immune from their scrutiny. Facebook eventually sent its chief technology officer, Mike Schroepfer, to the British parliamentary inquiry, but he failed to fully answer forty questions according to a subsequent statement by the committee. But what was perhaps most revealing about the performance was the lack of contrition on the part of the company. When Schroepfer was asked if Facebook's first instinct to send journalists legal threats was bullying behaviour, the Facebook CTO replied that 'my understanding is that this is common practice in the UK'. After being pressed by the incredulous MPs, Schroepfer acquiesced and finally apologised, saying that he was 'sorry that journalists feel we are attempting to prevent the truth coming out'.

Of all the individuals who could have been formally punished in this saga, it was sad for me to see that one of the only people to face a sanction was Darren Grimes, the twenty-two-year old Vote Leave intern. As frustrating as his situation was, the archaic legislation meant that he was personally liable for electoral offenses. The commission levied a £20,000 fine against him personally and referred his case to the police. He subsequently succeeded in an appeal against that finding, although further appeals may yet be brought by the Electoral Commission. The campaign, Vote Leave, was fined £61,000, part of which reflected their refusal to cooperate with the regulator. Vote Leave dropped their appeal and that sanction, at least, remains.

It was incredibly hard to watch what happened to Grimes, who had his life torn up over a scheme that others had orchestrated. We had hoped that he would come forward with Sanni, Gettleson and me, but Grimes defended the scheme until the very end. He panicked and broke down every time Sanni broached the topic, and did not want to accept that he had been used by the people he trusted. Grimes was set up to become their fall guy, and Vote Leave could not have asked for a better candidate. As much as he defended his old bosses' actions, Grimes was their captive victim. They transformed him from a talented, liberal and artistic student into a public shill for their alt-right causes, in exchange for help with legal fees.

Several weeks after the story went public, Shahmir Sanni was terminated from his job at the TaxPayers' Alliance, a think tank, after pressure from Conservative Party advisers. The alliance later admitted to his lawyers that they unlawfully fired Sanni in retaliation for what they called his 'philosophical belief in the sanctity of British democracy'. Although the question of Parkinson's job at 10 Downing Street was raised several times in Parliament, Parkinson kept his job and faced no consequences for using the press office of the prime minister to out his former intern as being gay. Before Theresa May stepped down as Prime Minister, she recommended Parkinson for a peerage in the House of Lords in her resignation honours list. Upon joining the Lords, Parkinson would be entitled to vote on laws and collect an allowance for the rest of his life. And Mark Gettleson, who provided evidence to authorities on both sides of the Atlantic, was pushed out of his new job at a mobile app company over reputational concerns about his whistleblowing.

In March 2018, just before the staff at Cambridge Analytica learned about the impending demise of their firm, Alexander Nix allegedly emptied £6 million from company accounts, preventing severance pay from being issued to its former staff. He later denied this at Parliament, saying that the withdrawn money was 'in exchange for unbooked services' and that he intended on paying some of it back. Nix was shunned by many of his former business partners and peers in the private clubs of Pall Mall, but, as a man of exceptional wealth, he could continue living off his inheritance in his mansion in London's Holland Park. Nothing much happened to him beyond some cringe-worthy public hearings in Parliament in which he blamed the 'global liberal media' for his company's demise.

After I came forward with the Cambridge Analytica story, Brittany Kaiser rebranded herself as a whistleblower and hired a PR manager to start booking interviews. She attended a parliamentary hearing in which she admitted to being involved in the Nigeria project, said that Cambridge Analytica likely retained Facebook data, and outlined her relationship with Julian Assange. (Later, it would emerge that she visited Assange in the Ecuadorian embassy in London.) Immediately after Kaiser's testimony concluded, Nix texted her, 'Well done Britt, it looked quite tough and you did ok. ;-).' The next day, she flew to New York and held a press conference to plug her new data project, which launched something called the Internet of Value Omniledger, apparently intended to unleash our 'data freedom'.

Like Kaiser, several other former executives from Cambridge Analytica went on to found their own data companies. CA's former head of product Matt Oczkowski founded a firm called Data Propria (Latin for 'Personal Data') and brought CA's chief data scientist David Wilkinson with him. The firm has stated that it will focus on targeting 'motivational behavioural triggers' and had already started work for the 2020 US presidential campaign of Donald Trump. Mark Turnbull, the former managing director of Cambridge Analytica, joined up with one of the firm's former associates, Ahmad Al-Khatib, to set up Auspex International, which they described as an 'ethically based' and 'boutique geopolitical consultancy'.

My biggest regret was Jeff Silvester. I can't even begin to put into words how maddening and disheartening it was for me to sit with the knowledge about what he and AIQ had done. He was my mentor when I was a teenager and the man who helped me enter politics in the first place. He had supported me, encouraged me and nurtured my talents so I could grow. And I just still cannot understand how he could have let himself continue working for something so wrong, so colonial, so illegal and so evil. I tried to talk to him, and I told him to be open with the *Guardian*, but I failed. He could have come clean. He could have cooperated with the investigations. He knew what AIQ had done was wrong. He knew that the effects of his work had profound consequences for the future of an entire nation and the rights of millions of people. Having to choose between a deep friendship and reporting a crime is torture, because no matter what you choose, you'll feel profound regret. But I had no choice but to betray him. On the day the *Guardian* sent out the right-to-reply letters to all of the accused parties, I agonised over what was happening the entire day, waiting to hear anything. When he received his letter, Silvester finally learned of the choice I had made, and he began to realise what was about to happen to him. His final text message to me was simply 'Wow'.

Walking into my first parliamentary hearing, to the sound of rapidly clicking cameras and shouted questions, I felt unexpectedly at ease. Allen sat behind me, occasionally passing me notes of legal advice. We had prepared for hours, going through the evidence, and I had the special protection of parliamentary privilege – meaning that nothing I said could be used in civil or criminal proceedings. The hearing caused a wave of legislative attention around the world, and the DCMS committee chair, Damian Collins, began organising international joint hearings among fifteen national parliaments. There were debates on the floor of the House of Commons and cross-party support for regulating social media. For a couple of months, it seemed as if Britain was leading the way in challenging the power of Silicon Valley.

But then, in October 2018, seven months after the Cambridge Analytica scandal rocked Facebook, the company announced that it was making a major hire: a new apologist in chief to world governments. Facebook's new global spin doctor was going to be Nick Clegg, the former leader of the Liberal Democrats and deputy prime minister of the United Kingdom – the same man I used to work for in my days at LDHQ. Ironically, it was Clegg who had once vowed that he would go to prison before registering in a pilot national identity database. But he was also the guy whose tenure as deputy prime minister became in effect a five-year apology tour after he broke a host of key promises in the coalition government. And the more I thought about it, the more the pairing seemed to be a match made in heaven. Both Zuckerberg and Clegg had built their careers on compromising their principles, both suffered catastrophic blows of public confidence after they ignored their promises to users or voters, and both stopped being cool in 2010. When Channel 4 asked me for comment on camera after Clegg's appointment was announced, all I could think to say was 'This is bullshit.' They aired the comment, albeit with a bleep.

On 24 May, 2019, Prime Minister Theresa May announced her intention to resign, triggering an internal leadership race within the Conservative Party. In the United Kingdom, if a prime minister resigns mid-term, the convention is that Her Majesty the Queen appoints the new leader of the governing party as the new prime minister without a general election. This means that the internal party back-roomers, donors and paid members of the party can bypass an election and choose among themselves who shall lead Britain. On 23 July, the members of the Conservative Party decided that the new prime minister would be Boris Johnson, the former foreign secretary and lead advocate for leaving the European Union without any negotiated exit deal (often referred to as a 'hard Brexit'). When forming his new government, Johnson appointed Dom Cummings, his former colleague from Vote Leave, to become one of his new senior advisers in 10 Downing Street. It did not seem to matter that Cummings was the director of a campaign that cheated during the very referendum Johnson was now using as the 'democratic' basis for leaving the European Union at almost any cost. Leaked documents from the Cabinet Office later revealed Johnson's advisers immediately started plans to suspend Parliament as one of the first acts of their new government. The suspension of Parliament would prevent MPs from scrutinising their designs for Brexit. This aversion to democratic scrutiny is unsurprising, however. Only a few months prior to his appointment, Cummings was found to be in contempt of Parliament after ignoring an order to appear before Parliament to answer questions about cheating and the dissemination of fake news in the EU referendum. Although Cummings is one of only a handful of people ever to be formally admonished by a unanimous vote of the House of Commons, the limits of parliamentary authority were tested, and it appears there were very few consequences for Cummings. He was even granted a parliamentary pass upon his appointment. And slated to join Cummings in the new Johnson government as a new special adviser to Her Majesty's Treasury was Matthew Elliott, the former chief executive of Vote Leave and co-founder of the TaxPayers' Alliance, the lobbying group that fired Sanni in retaliation for his whistleblowing. It looked like a Vote Leave takeover of the British government. During his first Prime Minister's Questions session in the House of Commons, Johnson was asked by opposition members about what was discussed in December 2016, when he met with Cambridge Analytica's CEO, Alexander Nix, when he was Britain's foreign secretary. His response was simply, 'I have no idea.'

Inside Cambridge Analytica, I saw what greed, power, racism and colonialism looks like up close. I saw how billionaires behave when they want to shape the world in their image. I saw the most bizarre, dark niches of our society. As a whistleblower, I saw what big companies will do to protect their profits. I saw the lengths to which people will go to cover up crimes that others committed for the sake of a convenient narrative. I saw flag-waving 'patriots' turn a blind eye to the defacement of the rule of law on the most important constitutional question of a generation. But I also saw all the people who cared and who fought back against a failing system. I saw journalists at the *Guardian*, *The New York Times* and Channel 4 all working to bear witness to the crimes committed by Cambridge Analytica and the incompetence of Facebook. I saw my brilliant lawyers outmanoeuver every threat that was thrown my way. I saw the kindness of people who came to support me and asked for nothing in return. I saw the tiny Information Commissioner's Office, based in the parish town of Wilmslow, use what powers it could to take on an American technology giant – eventually issuing Facebook the maximum fine allowable in law for data breaches.

And I saw members of Congress who were concerned and eager to learn about the brave new world we now find ourselves in. As I left the House Intelligence Committee hearing, emerging from the SCIF with my lawyers and Sanni, I shook hands with the members of the committee and was walked to the security entrance by Congressman Adam Schiff and his aides. They were gracious, and they thanked me for flying to America to help them understand not only Cambridge Analytica but the emerging risks to American elections posed by social media platforms. It would be the last of my testimonies in the United States, but everything felt far from resolved.

On 24 July, 2019, the Federal Trade Commission levied a record $5 billion civil penalty against Facebook, and the same day the Securities and Exchange Commission issued notice of an additional $100 million fine. The regulators found that not only did Facebook fail to protect users' privacy, the company misled the public and journalists by issuing false statements that it had seen no evidence of wrongdoing when it in fact had. The fine was one of the largest imposed by the US government for any violation. In fact, this was the largest ever fine issued to an American company for violating consumers' privacy rights, and was twenty times greater than the largest privacy or data security penalty ever imposed worldwide. However, it was nonetheless seen by investors as good news. The news actually increased Facebook's share value by 3.6 per cent, with the market tacitly recognising that even the law cannot stop the growth of these technology giants.

I would be lying if I didn't admit that I am far more cynical now than before I started this journey. But it hasn't made me more resigned. If anything, it has made me even more radical. I used to believe that the systems we have broadly work. I used to think that there was someone waiting with a plan who could solve a problem like Cambridge Analytica. I was wrong. Our system is broken, our laws don't work, our regulators are weak, our governments don't understand what's happening, and our technology is usurping our democracy.

So I had to learn to find my voice in order to speak up about what I saw was happening. I am hopeful, because I have seen what happens when we find our voices. When the *Guardian* took on this story, many journalists saw it as a series of conspiracy theories. The tech bros of Silicon Valley laughed at the notion that they should be subjected to any scrutiny. Politicos in DC and Westminster called the story *niche*. It took the persistence of a team of women at the *Guardian*'s Arts & Culture section and its Sunday paper, the *Observer,* where the blockbuster story appeared. It took the attention of the women who led the investigations at the Information Commissioner's Office and the

Electoral Commission. And it took two immigrant queer whistleblowers backed by a steadfast woman lawyer. This story took the leadership of dedicated women, immigrants and queers to ignite a public awakening about the discreet colonising power of Silicon Valley and the digital technologies they have created to surround us. We all persisted in raising our voices until the world could finally see what we saw.

Growing up queer, you learn early in life that your existence is outside the norm. We incubate ourselves inside a closet, remaining unknown, and hide our truth until it becomes unbearable. Living in a closet is painful. It is an act of emotional violence we inflict upon ourselves so as not to discomfort those around us. Queers understand systems of power intimately, and coming out is our transformative act of truth telling. In coming out, we realise the power of speaking our truth to those who may not want to hear it. We reject their comfort and make them listen. Why do so many gays blow whistles at Pride? To get your attention. To announce that we will no longer hide ourselves. To defy hegemonies of the powerful. And, like so many queers who came before me, I had to accept my own truth and come to terms with my inevitable failure to ever become society's notion of a perfect man.

I am a queer whistleblower, and this was my second coming out. Subjecting me to covenants of nondisclosure, I was forced into a new closet, to live in hiding with my uncomfortable knowledge and objectionable truths. I lived my life for two years with a personalised *don't ask, don't tell* policy imposed upon me by powerful companies. If I hoped to avoid any consequences, I was forbidden to reveal myself to others, and I became their little secret. But like other out queers, I am a truth teller, and I chose to be indiscreet with those uncomfortable truths, to stop hiding, to stop being their secret, to face the consequences before me, and to shout out to the world what I know.

The closet is not a literal space; it is a social structure that we as queer people internalise and conform to. The closet is a container whose boundaries are imposed by others who want to control how you behave and present yourself. The closet is invisible, and it is placed upon you by default, never by choice, for others to create a more palatable version of who you are – for *their* benefit, *not yours*. Growing up in a closet means incrementally learning how to *pass* in society – which movements, tones, expressions, perspectives, or uttered desires transgress the norms of those social boundaries imposed upon you. Queer kids learn, little by little, how to restrain their behaviour until it becomes almost second nature, until they pass. So incremental are these changes that sometimes you do not even notice how much you have changed your behaviour until, one day, you decide to leave that closet. And part of coming out is coming to terms with how much of you has been constructed for you inside that closet, and it can be painful to realise how much of who you once were was imposed upon you without your awareness or consent. The closet is a place of acquiescing to society in exchange for *passing*, but it is also a place where rage builds as those boundaries and definitions slowly suffocate you until you cannot bear to remain inside that prison.

Coming out is our rejection of the definitions that have been imposed upon us by someone else. The ability to define our identities is extremely powerful, and whether the threats to that power take the form of a social closet or an algorithmic one, we must resist anyone or anything that seeks the power to define or classify who we are for their benefit. Silicon Valley risks creating a new hegemony of identity through its construction of these personalised spaces for each person. And these spaces are nothing but a new closet to define our identities, expressions and behaviours. In harvesting and processing your data self, algorithms make decisions on how to define you, how to classify you, what you should notice and who should notice you. But there is a fine line between an algorithm defining you in order to represent *who you really are* and an algorithm defining you to create a self-fulfilling prophecy of *who it thinks you should become*.

People are *already* morphing themselves to fit a machine's idea of who they should be. Some of us are curating ourselves on social media to increase our follower engagement, to the point that who we really are and how we present online become confused and conflated. And when those followers see enough of these curated identities, some of them begin to hate who they are or how they look, and they starve their bodies to conform to a new standard that now surrounds them.

Others click on links recommended to them by algorithms, engaging with that content, and get drawn further and further down the rabbit hole of personalisation until their worldview changes without their realising it. What we buy online is now curated based on a profile of us, *defined by something else.* Our worthiness as job, insurance, credit, or mortgage applicants is now based on a profile of us, *defined by something else.* The shows we watch and the music we discover are now preselected based on a profile of us, *defined by something else.* As we move towards the inevitable merger of our physical and digital worlds, more and more of our lives will start to become defined not by us but *by something else.* And so, if we are ever to resist our future lives being *defined by something else,* we may all need to come out of our closets before someone or something locks us inside.

ON 23 MAY, 2019, I woke up at 6 a.m., unusually early for me. My room was bright and warming up, the sunrise peeking through my curtains. I hate getting up early, so I stared at the ceiling for a bit before glancing out the window to see life emerging on the street. A guy I had been seeing stayed the night, so I had to slip out of bed carefully in order not to make a sound. It was polling day in Britain, in what was potentially the last-ever European Parliament election. My polling card said polls would open at 7 a.m., so I wanted to sneak out to run to the local community centre where voting in my local ward was taking place.

Taking slightly exaggerated steps to silently glide over to my dresser, I grabbed my jeans and a T-shirt, lying in a heap on the floor. The shirt was a gift from the English designer Katharine Hamnett. Soft black cotton with bold white letters, it simply read, SECOND REFERENDUM NOW! *If I wear anything today, it should be this T-shirt,* I thought. I reached over into my drawer to pull out my phone, and once it regained signal, it began buzzing with messages.

Oh shit, I thought. I turned back to see I had woken him up. Groaning into a pillow, he asked why I was up so early and I simply said because I want to go vote. He sat up and smirked, rolling his eyes, asking if today was like Christmas for *people like me.* I told him no, that I wanted to go early, before the party poll watchers show up and start tallying who is voting. I didn't want to get into another fight with UKIP or Brexiteers. I have been called a traitor and pushed into the streets, but I did not want to be stopped from voting.

It did not feel like Christmas, and it wasn't exciting at all. It was a sad day, because I knew in my heart that I wasn't going to be taking part in a *real* election – it was all part of a final performance before Britain was scheduled to leave the European Union. Despite the Electoral Commission's ruling against Vote Leave, an ongoing National Crime Agency investigation, testimonies at Parliament, and a week-long exposé in the *Guardian* about the cover-up inside Downing Street, the government was nonetheless determined to exit the European Union with a mandate won through cheating and fraud.

My postbox was filled with leaflets and literature. I was half expecting to receive something mad from Arron Banks or Leave.EU, like a Brexit leaflet rolled into a Russian vodka bottle, as they were so fond of trolling me and *Guardian* journalist Carole Cadwalladr. But no, it was just regular leaflets. Greens. Lib Dems. UKIP. Nothing from the Tories or Labour, for some reason. I opened up the Lib Dem one and I thought about what data they were using now and whether they had targeted me with a message. It didn't look like it. It was just another crap leaflet.

I looked up at the security camera watching me in the lobby and left. I set out, walking through a couple of streets in my neighbourhood. Old Georgian row houses interspersed with the occasional block of flats. It was extremely bright and sunny. The morning air was fresh and invigorating. I turned onto a high street, where the shops were not yet open, save for a local coffee shop. I walked in and ordered a coffee with a splash of soy milk. As I waited, I looked at everyone in the café, standing and looking at their phones, all scrolling, following and engaging with content. I stood beside them, but they were all off in their own digital worlds. To be honest, I used to do the same thing before my ban. But without social media, aside from a Twitter account I barely use, I have found myself scrolling less, posting less and taking fewer photos of things. I no longer spend hours being alone together with other people through my screen. I may live outside these digital worlds, but at least I have come to be more present in this world. After grabbing my coffee, I left and walked down a tree-lined street before reaching the community centre. Tied to the trees were large white placards with black letters that read POLLING STATION. I kept my distance and peered around, but no one from any of the parties was loitering outside yet. So I walked inside and followed the signs down a corridor and into a simple, unadorned room scattered with cardboard voting booths and tiny pencils without erasers.

The polling station clerk looked at me and asked for my name. She flipped through the paper list and took a pencil to cross it out. That was it – no IDs, no electronics. She handed me what seemed like a metre-long ballot for the election of London's delegation of members of the European Parliament. The paper was only slightly thicker than newspaper, but as I held it, I thought about how physical the act of voting seems, and yet so much sophisticated activity online leads up to this simple act of crossing an X on a thin piece of paper. I dropped the ballot into the ballot box and hoped it would not be the last time.

EPILOGUE

On Regulation:
A Note to Legislators

If we are to prevent another Cambridge Analytica from attacking our civil institutions, we must seek to address the flawed environment in which it incubated. For too long the congresses and parliaments of the world have fallen for a mistaken view that somehow 'the law cannot keep up with technology'. The technology sector loves to parrot this idea, as it tends to make legislators feel too stupid or out of touch to challenge their power. But the law *can* keep up with technology, just as it has with medicines, civil engineering, food standards, energy and countless other highly technical fields. Legislators do not need to understand the chemistry of molecular isomers inside a new cancer drug to create effective drug review processes, nor do they need to know about the conductivity of copper in high-voltage wiring to create effective insulation safety standards. We do not expect our legislators to have expert technical knowledge in any other sector because we devolve technical oversight responsibility to regulators. Regulation works because we trust people who know better than we do to investigate industries and innovations as the guardians of public safety. 'Regulation' may be one of the least sexy words, evoking an image of faceless jobsworths with their treasured checklists, and we will always argue about the details of their imperfect rules, but nonetheless safety regulation generally works. When you buy food in the grocery store or visit your doctor or step onto an airplane and hurtle thousands of feet in the air, do you feel safe? Most would say yes. Do you ever feel like you need to think about the chemistry or engineering of any of it? Probably not.

Tech companies should not be allowed to *move fast and break things*. Roads have speed limits for a reason: to slow things down for the safety of people. A pharmaceutical lab or an aerospace company cannot bring new innovations to market without first passing safety and efficacy standards, so why should digital systems be released without any scrutiny? Why should we allow Big Tech to conduct scaled human experiments, only to realise that they become too big a problem to manage? We have seen radicalisation, mass shootings, ethnic cleansing, eating disorders, changes in sleep patterns and scaled assaults on our democracy, all directly influenced by social media. These may be intangible ecosystems, but the harms are not intangible for victims.

Scale is the elephant in the room. When Silicon Valley executives excuse themselves and say their platform's scale is so big that it's really hard to prevent mass shootings from being broadcast or ethnic cleansing from being incited on their platforms, this is not an excuse – they are implicitly acknowledging that what they have created is too big for them to manage on their own. And yet, they also implicitly believe that their right to profit from these systems outweighs the social costs others bear. So when companies like Facebook say, 'We have heard feedback that we must do more,' as they did when their platform was used to live-broadcast mass shootings in New Zealand, we should ask them a question: If these problems are too big for you to solve on the fly, why should you be allowed to release untested products before you understand their potential consequences *for* society?

We need new rules to help create a healthy friction on the internet, like speed bumps, to ensure safety in new technologies and ecosystems. I am not an expert on regulation, nor do I profess to know all the answers, so do not take these words as gospel. This should be a conversation that the wider community takes part in. But I would like to offer some ideas for consideration – at the very least to provoke thought. Some of these ideas may work, others may not, but we have got to start thinking about this hard problem. Technology is powerful, and it has the potential to lift up humanity in so many ways. But that power needs to be focused on constructive endeavors. With that, here are some ideas to help you consider how to move forward:

1. A Building Code for the Internet

The history of building codes stretches back to the year 64 CE, when Nero restricted housing height, street width and public water supplies after a devastating fire ravaged Rome for nine days. Though a fire in 1631 prompted Boston to ban wooden chimneys and thatched roofs, the first modern building code emerged out of the devastating carnage of the Great Fire of London, in 1666. As in Boston, London houses had been densely constructed from timber and thatch, which allowed the fire to spread rapidly over four days. It destroyed 13,200 homes, eighty-four churches, and nearly all of the city's government buildings. Afterward, King Charles II declared that no one shall 'erect any House or Building, great or small, but of Brick, or Stone'. His declaration also widened thoroughfares to stop future fires from spreading from one side of the street to the other. After other historic fires in the nineteenth century, many cities followed suit, and eventually public surveyors were tasked with inspections and ensuring that the construction of private property was safe for the inhabitants and for the public at large. New rules emerged, and eventually the notion of *public safety* became an overarching principle that could override unsafe or unproven building designs, regardless of the desires of property owners or even the consent of inhabitants. A platform like Facebook has been burning for years with its own disasters – Cambridge Analytica, Russian interference, Myanmar's ethnic cleansing, New Zealand's mass shootings – and, as with the reforms after the Great Fire, we must begin to look beyond policy, to the underlying architectural issues that threaten our social harmony and citizens' well-being.

The internet contains countless types of architectures that people interact with on a daily and sometimes hourly basis. And as we merge the digital world with the physical world, these digital architectures will impact our lives more and more. Privacy is a fundamental human right and should be valued as such. However, too often privacy is eviscerated through the bare performance of clicking 'accept' to an indecipherable set of terms and conditions. This *consent-washing* has continually allowed large tech platforms to defend their manipulative practices through the disingenuous language of 'consumer choice'. This positions our frame of thinking away from the design – and the designers – of these flawed architectures, and towards an unhelpful focus on the activity of a user who does not have understanding or control over the system's design. We do not let people 'opt in' to buildings that have faulty wiring or lack fire exits. That would be unsafe – and no terms and conditions pasted on a door would let any architect get away with building dangerous spaces. Why should the engineers and architects of software and online platforms be any different?

In this light, consent should not be the sole basis of a platform's ability to operate a feature that engages the fundamental rights of users. In following the Canadian and European approach of treating privacy as an engineering and design issue – a framework called 'privacy by design' – we should extend this principle to create an entire engineering code: a building code for the internet. This would include new principles beyond privacy, to include respect for the agency and integrity of end users. Such a code would create a new principle – agency by design – to require that platforms use *choice-enhancing* design. This principle would also ban dark pattern designs, which are common design patterns that deliberately confuse, deceive or manipulate users into agreeing to a feature or behaving in a certain way. Agency by design would also require *proportionality of effects,* wherein the effect of the technology on the user is proportional to the purpose and benefit to the user. In other words, there would be a *prohibition on undue influence* in platform design, where there are enduring and disproportionate effects, such as addictive designs or consequential mental health issues.

As with traditional building codes, the *harm avoidance principle* would be a central feature in such a digital building code. This would require platforms and applications to conduct *abusability audits* and safety testing *prior* to releasing or scaling a product or feature. The burden would rest with tech companies to prove that their products are safe for scaled use in the public. As such, using the public in live scaled experiments with untested new features would be prohibited, and citizens could no longer be used as guinea pigs. This would help prevent cases like Myanmar, where there was no prior consideration from Facebook on how features could ignite violence in regions of ethnic conflict.

2. A Code of Ethics for Software Engineers

If your child was lost and needed help, whom would you want them to turn to for help? Perhaps a doctor? Or maybe a teacher? What about a cryptocurrency trader or gaming app developer? Our society esteems certain professions with a trustworthy status – doctors, lawyers, nurses, teachers, architects and the like – in large part because their work requires them to follow codes of ethics and laws that govern safety. The special place these professions have in our society means that we demand a higher standard of professional conduct and duties of care. As a result, statutory bodies in many countries regulate and enforce ethical conduct of these professions. For society to function, we must be able to trust our doctors or lawyers to always act in our interests, and that the bridges and buildings we use every day have been constructed to code and with competence. In these regulated professions, unethical behaviour can bring dire consequences for those who transgress boundaries set by the profession – ranging from fines and public shaming to temporary suspensions or even permanent bans for more egregious offenders.

Software, AI and digital ecosystems now permeate our lives, and yet those who make the devices and programs we use every single day are not obligated by any federal statute or enforceable code to give due consideration to the ethical impacts to users or society at large. As a profession, software engineering has a serious ethics problem that needs to be addressed. Tech companies do not magically create problematic or dangerous platforms out of thin air – there are people inside these companies who build these technologies. But there is an obvious problem: software engineers and data scientists do not have any skin in the game. If an engineer's employer instructs her or him to create systems that are manipulative, ethically dubious or recklessly implemented, without consideration for user safety, there are no requirements to refuse. Currently, such a refusal to act unethically creates a risk to the employed engineer of repercussions or termination. Even if the unethical design later is found to run afoul of a regulation, the company can absorb liability and pay fines, and there are no professional consequences for the engineers who built the technology, as there would be in the case of a doctor or lawyer who commits a serious breach of professional ethics. This is a perverse incentive that does not exist in other professions. If an employer asked a lawyer or nurse to act unethically, they would be obligated to refuse or face losing their professional license. In other words, they have skin in the game to challenge their employer.

If we as software engineers and data scientists are to call ourselves professionals worthy of the esteem and high salaries we command, there must be a corresponding duty for us to act ethically. Regulations on tech companies will not be nearly as effective as they could be if we do not start by putting skin in the game for people inside these companies. We need to put the onus on engineers to start giving a damn about what they build. An afternoon employee workshop or a semester course on ethics is a wholly insufficient solution for addressing the problems we now face from emerging technologies. We cannot continue down the path we are on, in which technological paternalism and the insulated bro-topias of Silicon Valley create a breed of dangerous masters who do not consider the harm that their work has the potential to inflict.

We need a professional code that is backed by a statutory body, as is the case with civil engineers and architects in many jurisdictions, where there are actual consequences for software engineers or data scientists who use their talents and know-how to construct dangerous, manipulative or otherwise unethical technologies. This code should not have loose aspirational language; rather, what is acceptable and unacceptable should be articulated in a clear, specific and definitive way. There should be a requirement to respect the autonomy of users, to identify and document risks, and to subject code to scrutiny and review. Such a code should also include a requirement to consider the impact of their work on vulnerable populations, including any disproportionate impact on users of different races, genders, abilities, sexualities or other protected groups. And if, upon due consideration, an employer's request to build a feature is deemed to be unethical by the engineer, there should be a *duty to refuse* and a *duty to report*, where failure to do so would result in serious professional consequences. Those who refuse and report must also be protected by law from retaliation from their employer.

Out of all the possible types of regulation, a statutory code for software engineers is probably what would prevent the most harm, as it would force the builders themselves to consider their work *before anything is released* to the public and not shirk moral responsibility by simply following orders. Technology often reflects an embodiment of our values, so instilling a culture of ethics is vital if we as a society are to increasingly depend on the creations of software engineers. If held properly accountable, software engineers could become our best line of defence against the future abuses of technology. And, as software engineers, we should all aspire to earn the public's trust in our work as we build the new architectures of our societies.

3. Internet Utilities and the Public Interest

Utilities are traditionally physical networks said to be 'affected with the public interest'. Their existence is unique in the marketplace in that their infrastructure is so elemental to the functioning of commerce and society that we allow them to operate differently from typical companies. Utilities are often by necessity a form of natural monopoly. In a marketplace, balanced competition typically results in innovation, better quality and reduced prices for consumers. But in certain sectors, such as energy, water or roads, it makes no sense to build competing power lines, pipelines or subways to the same places, as it would result in massive redundancy and increased costs borne by consumers. With the increased efficiencies of a single supplier of a service comes the risk of undue influence and power – consumers unable to switch to new power lines, pipelines or subways could be held hostage by unscrupulous firms.

On the internet, there are clearly extremely dominant market players. Google accounts for more than 90 per cent of all search traffic, and almost 70 per cent of adults active on social media use Facebook. But this does not make them universal infrastructure per se. When tech platforms suffer an outage, we can survive and cope for longer (albeit not indefinitely) than we would if the same thing happened with electricity. On the infrequent occasions when Google's search engine has failed, users coped by moving to other, lesser-known search engines until Google patched its problem. There are also cycles of popularity for large internet players that are not found in physical infrastructure. MySpace was once the preeminent social media platform, before Facebook crushed it, whereas we rarely if ever encounter market cycles with water or electric companies.

That said, the internet's dominant players do share things in common with physical utilities. Like physical utilities, these architectures often serve as a de facto backbone of commerce and society, where their existence has become a given of day-to-day life. Businesses have come to passively rely upon the existence of Google's search engine for their workforces, for example. And this is not a bad thing. Search engines and social media benefit from network effects, where the more people use the service, the more useful the service becomes. As with physical utilities, scale can create a huge benefit to the consumer, and we do not want to impede this public benefit. However, as with other natural monopolies, the same kinds of risks threaten consumers. And it is these potential harms that we must account for in a new set of rules.

So, in full recognition that there are essential differences between the internet and physical infrastructure, I will use 'internet utilities' as a term of convenience to mean something *similar to but different from* a traditional utility: An 'internet utility' is *a service, application or platform whose presence has become so dominant on the internet that it becomes affected with the public interest by the very nature of its own scale.* The regulation of internet utilities should recognise the special place they hold in society and commerce and impose a higher standard of care toward users. These regulations should take the form of statutory duties, with penalties benchmarked to annual profits as a way to stop the current situation, in which regulatory infractions are negotiated and accounted for as a cost of doing business.

In the same way we do not penalise the scale of electric companies, the scale of internet utilities should not be penalised where there are network effects of genuine social benefit. In other words, this is not about breaking up large technology companies; this is about making them accountable. However, in exchange for maintaining their scale, internet utilities should be required to act proactively as responsible stewards of what eventually evolve into *our* digital commons. They must be made to understand that scale evokes innate public interests that in some cases will, by necessity, supersede their private interests in making profit. As with other utilities, this must include compliance with higher user safety standards specific to software applications and a *new code of digital consumer rights*. These new digital consumer rights should serve as the basis of universal terms and conditions, in that the interests of internet users are properly accounted for in areas where technology companies have continuously failed.

4. Public Stewardship of the Digital Commons

The unrestricted power of these internet utilities to impact our public discourse, social cohesion and mental health, whether intentionally or through incompetence and neglect, must also be subject to public accountability. A new *digital regulatory agency* should be established to enforce this new digital regulatory framework with statutory sanctioning powers. In particular, these agencies should contain technically competent ombudsmen empowered with rights to conduct proactive technical audits of platforms on behalf of the public. We should also use market-based reinforcement mechanisms, such as the requirement of internet utilities to hold insurance for harms incurred from data misuse. In requiring insurance for data breaches, linked to the market-rate value of that data, we could create corrective financial pressure to do better.

We have seen the value of personal data create entirely new business models and huge profits for social media companies. Platforms such as Facebook have vigorously argued that they are a 'free' service, and if consumers do not have to pay for the service, the platform cannot be complicit in anticompetitive practice. However, this argument requires one to accept that the exchange of personal data for use of a platform is not an exchange of value, when it plainly is. There are entire marketplaces that valuate, sell and license personal data. The flaw in current antitrust approaches for large tech companies is that the value of the consumer's data has not been properly accounted for by regulators.

If we actually considered the rising value in the personal data provided by consumers to platforms, we'd conclude that consumers have continuously been ripped off by these companies, which are not proportionately increasing their platform's value to consumers. In this light, consumers are giving more value via their data to these dominant platforms without receiving reciprocal benefits. There may be an argument in America's current antitrust laws that on this basis, the data exchange has been costing consumers more. However, even if this were the case, this is an overly narrow test of what is fair and right for consumers. Instead, were we to create a new classification of internet utilities, we could use a broader *public interest test* for the operation, growth and mergers-and-acquisitions activity of these firms.

However, unlike physical utilities, social media and search engines are not so essential as to be irreplaceable, so regulations should also account for the healthy benefits of industry evolution. We want to avoid regulations that entrench the position of currently dominant internet utilities at the expense of newer and better offerings. But we also need to reject the notion that any regulation of scaled giants would somehow impede new challengers. To follow this logic would mean that safety and environmental regulation of the oil sector would inhibit the emergence of renewable energy in the future, which makes no sense. And if we are concerned about the inhibition of market evolution, then we could require internet utilities to share their dominant infrastructure with smaller, competing challengers, to improve consumer choice in the same way dominant telecom companies share communications infrastructure with smaller players. Safety and conduct standards of existing large players are not mutually incompatible with the continuation of technological evolution. In this light, *principle-based* rather than *technology-based* regulation should be created so that we are careful not to embed old technologies or outdated business models into regulatory codes.

Thanks. And good luck.

ACKNOWLEDGMENTS

SO OFTEN WHISTLEBLOWERS ARE POSITIONED AS LONE DAVIDS on a mission to singlehandedly take on Goliath. But in my case, I was never alone. There were so many people without whom none of this would have been possible. From lawyers to journalists, sisters to taxi drivers – many people have contributed greatly to this story, and I am so utterly grateful for their advice, resilience, patience and tenacity. I want to especially acknowledge all of the women who supported me in this journey. It was women who made this story possible.

THE LAWYERS

For defending me no matter the odds, and for being the coolest lawyer someone could ever have, I would like to first thank my brilliant lead lawyer, Tamsin Allen. Tamsin, you helped me before anyone knew who I was, or what Cambridge Analytica did. You made it possible for me to take on some of the most powerful people and companies in the world. When I had to travel to Washington, DC, to testify before the US House Intelligence Committee, I learned three things about you. First, you are afraid of flying. Second, literally nothing else seems to faze you. Third, even after a transatlantic flight, bouts of jet lag, and sitting with me for more than five intense hours of congressional hearings, you were still able to outdance Jennifer Lopez at the TIME 100 gala later that evening.

There were so many brilliant lawyers who worked tirelessly behind the scenes to protect me and help make this story possible. Adam Kaufmann, Eric Lewis, Tara Plochocki and my whole US legal team at Lewis Baach Kaufmann Middlemiss PLLC – thank you for so bullishly taking on my case, effortlessly handling its multijurisdictional complexity, and helping me come through this process unscathed. Your counsel was imperative to keeping me calm and collected in such a chaotic time. In Britain, I was also supported by Tamsin's brilliant colleagues at Bindmans LLP, including Mike Schwarz and Salima Budhani, and a small battalion of barristers from Matrix Chambers, including Gavin Millar QC, Clare Montgomery QC, Helen Mountfield QC, and Ben Silverstone and Jessica Simor QC. Martin Soames and Erica Henshilwood from Simons Muirhead & Burton LLP helped me greatly when I was still working with the *Guardian* anonymously, and their early advice laid the foundation for the rest of the story. You are all incredible lawyers, and I am here today, safe and sound, because of your work.

The Whistleblowers

Mark Gettleson and Shahmir Sanni, thank you for the deeply personal sacrifices you have both made and for sharing this crazy journey with me. You both experienced deeply unfair retaliation, but you nonetheless chose to blow the whistle. Mark, from the moment I met you all those years ago, there have been few men who have matched your eloquence, humour, empathy and intelligence. Shahmir, thank you for standing by my side since we started on our whistleblowing journeys together, and for speaking truth to power. We've been through hell and back together and I'm so utterly proud to call you both my friends. And to the several other whistleblowers who wish to remain anonymous, thank you for helping. Even if the world does not know your contribution, you have made a huge difference.

The Journalists

Carole Cadwalladr, thank you for believing me – and for believing *in* me. I knew from the moment I met you that you were one of the few people who could tell this story to the world in a way that makes people take note. You woke up the world and shook giants. I may have had the pink hair, but you were the one who wielded the pen. You kept going despite an unrelenting stream of abuse and threats from the alt-right, private intelligence firms and Silicon Valley tech bros. You took me on for no other reason than a sincere dedication to the greater good and you deserve every accolade for your brilliant journalism.

Sarah Donaldson and Emma Graham-Harrison, thank you for your pivotal role in telling this story to the world. Your work alongside Carole's is largely why I can so confidently claim that I would not be where I am today without the women involved. The *Guardian* and *Observer* are lucky to have you both. And of course, thank you, Paul Webster, John Mulholland and Gillian Phillips, for steadfastly defending this story in the face of billionaires, tech giants, angry White House officials, intelligence agencies and a cornucopia of almost daily legal threats. Matthew Rosenberg, Nicholas Confessore, Gabriel Dance, Danny Hakim, David Kirkpatrick and *The New York Times*, thank you for bringing this story to America in a way no one else could, and for the huge impact of your role in holding Facebook and other Silicon Valley giants to account. Job Rabkin, Ben de Pear, and Channel 4 News, thank you for daring to go undercover at huge risk, and for breaking this story to a television audience when others would not. Your footage showed the world the true depth of Cambridge Analytica's nefarious operation in the firm's own chilling words.

THE PARLIAMENTARIANS

Alistair Carmichael MP, thank you for your unwavering allyship and counsel over the years, for the late-night talks in your office, and for cultivating my palate for Scotch whisky during a stressful time. Your help before this story went public was invaluable. For nothing in return, you took risks and used your detailed knowledge of Parliament to help protect me and several other whistleblowers. This allowed evidence of significant public interest to be preserved and published. Damian Collins MP and the entire Digital, Culture, Media and Sport Committee at the British Parliament, thank you for being some of the loudest voices for holding Silicon Valley to account. Your nonpartisan collaboration on the inquiry into disinformation and 'fake news' put the public interest first, and you all set a shining example of how politics should be done. By working together, your committee took on the giants of Silicon Valley and pulled together support for legislative action from around the world. And Damian, as a bleeding-heart liberal I never thought I would ever say this, but you showed me that – *maybe* – some Tories really can be cool.

The Unsung Heroes

Thank you to my parents, Kevin and Joan, for your unconditional love, encouragement and wisdom, and to my two sisters, Jaimie and Lauren, for dropping everything to help when things got chaotic, for letting me vent my stress, and for keeping my fridge stocked with food. And thank you to everyone else who helped uncover and tell this story. In particular, I would like to thank Lord Strasburger (for your discreet but immeasurable assistance behind the scenes); Peter Jukes (for all your encouragement and a brilliant launch for the story); Marc Silver (for your stunning film and hours-long inspiring conversations); Jess Search (for your sage advice and nurturing my queerness); Kyle Taylor (for your passionate campaigning); Elizabeth Denham, Michael McEvoy, and the entire UK Information Commissioner's Office (for putting data rights on the map); Representative Adam Schiff and the staff of the US House Intelligence Committee (for all the unseen work you do); Glenn Simpson and Fusion GPS (for your brilliant investigative work); Ken Strasma (for sparking my interest in data); Dr Keith Martin PC (for nurturing my independent spirit); Jeff Silvester (for mentoring my younger self, despite everything that happened later); Tom Brookes (for your support throughout); David Carroll and Paul-Olivier Dehaye (for your persistence in defending our data rights); Dr Emma Briant (for uncovering critical evidence); Harry Davies, Ann Marlowe and Wendy Siegelman (for your early investigative work); my former academic supervisor Dr Carolyn Mair (for reviewing this book and teaching me so much about psychology, data and culture); and Professor Shoshana Zuboff (whose work on surveillance capitalism helped me refine so many ideas). Perhaps most important, I want to recognise the hundreds of thousands of people who shared this story, called their representatives, marched in protests, held up placards and sent me encouraging messages – there are so many people I have never even met who have passionately had my back throughout this journey.

This Book

And last, I would like to thank my two brilliant book collaborators, Lisa Dickey and Gareth Cook; my editor at Random House, Mark Warren; my literary agents at William Morris Endeavor, Jay Mandel and Jennifer Rudolph Walsh; Kelsey Kudak for fact-checking this book; and my entertainment lawyer, Jared Bloch. You all guided me through writing my first book, pushed me to put pen to paper, helped me distil the essence of this story, edited out my nonsense and curbed my more discursive tendencies.